Studies in Big Data

Volume 71

The series "Studies in Big Data" (SBD) publishes new developments and advances in the various areas of Big Data- quickly and with a high quality. The intent is to cover the theory, research, development, and applications of Big Data, as embedded in the fields of engineering, computer science, physics, economics and life sciences. The books of the series refer to the analysis and understanding of large, complex, and/or distributed data sets generated from recent digital sources coming from sensors or other physical instruments as well as simulations, crowd sourcing, social networks or other internet transactions, such as emails or video click streams and other. The series contains monographs, lecture notes and edited volumes in Big Data spanning the areas of computational intelligence including neural networks, evolutionary computation, soft computing, fuzzy systems, as well as artificial intelligence, data mining, modern statistics and Operations research, as well as self-organizing systems. Of particular value to both the contributors and the readership are the short publication timeframe and the world-wide distribution, which enable both wide and rapid dissemination of research output.

** Indexing: The books of this series are submitted to ISI Web of Science, DBLP, Ulrichs, MathSciNet, Current Mathematical Publications, Mathematical Reviews, Zentralblatt Math: MetaPress and Springerlink.

More information about this series at http://www.springer.com/series/11970

Mohammad Ayoub Khan ·
Mohammad Tabrez Quasim ·
Fahad Algarni · Abdullah Alharthi
Editors

Decentralised Internet of Things

A Blockchain Perspective

Editors
Mohammad Ayoub Khan
College of Computing and Information Technology
University of Bisha
Bisha, Saudi Arabia

Mohammad Tabrez Quasim
College of Computing and Information Technology
University of Bisha
Bisha, Saudi Arabia

Fahad Algarni
College of Computing and Information Technology
University of Bisha
Bisha, Saudi Arabia

Abdullah Alharthi
College of Computing and Information Technology
University of Bisha
Bisha, Saudi Arabia

ISSN 2197-6503 ISSN 2197-6511 (electronic)
Studies in Big Data
ISBN 978-3-030-38679-5 ISBN 978-3-030-38677-1 (eBook)
https://doi.org/10.1007/978-3-030-38677-1

This Springer imprint is published by the registered company Springer Nature Switzerland AG
The registered company address is: Gewerbestrasse 11, 6330 Cham, Switzerland

This book is dedicated to my loving father Dr. Mohammad Quasim who inspired me to live with purpose and honesty. He lost his battle with life while I was completing this book. I know Dad, now you cannot read this book because you have left us in sorrow. I pray to Almighty God to give me strength to follow your footprints and achieve success and prosperity in life. I believe you are always with me even if you are far away

Mohammad Tabrez Quasim

Foreword

One of the greatest innovations of this century is various developments in the information and communications (ICT) field, particularly Internet. ICT has basically changed the way we live, operate and interact with the world. The role of the Internet has gradually changed a lot because of the development in various communication technologies. Now, a day's billions of people and physical devices are connected via the Internet. The machines, people, smart objects, surrounding space and platforms are connected with wireless/wired sensors is an example of Internet of Things (IoT). The Internet of Things is a key enabler for many contemporary digital transformation domains such as smart cities, transportation systems, energy systems, healthcare and others.

This book is an excellent collection of chapters dealing with the conceptual knowledge of the latest tools and methodologies of decentralised Internet of Things and variety of real-world challenges. The topics cover the state-of-the art and future perspectives of Internet of Things technologies, where industry experts, researchers and academics had shared ideas and experiences surrounding frontier technologies, breakthrough and innovative solutions and applications. Each chapter presents the reader with an in-depth investigation regarding the role of blockchain and decentralised computing in the context of IoT setting.

The reader will find in this book a variety of decentralised techniques and applications in an IoT setting. Most aspects of various decentralised computing, blockchain and Internet of Things are critically examined in this book. The work in this book can be seen as a continuation of the quest for building decentralised IoT systems.

I would like to congratulate the editors for the timely edition of this valuable book, and I hope that researchers and others will be inspired by many of the ideas presented and will use and build upon them.

Prof. (Dr.) Ahmed Bin Hamid Nagadi
President, University of Bisha
Bisha, Kingdom of Saudi Arabia
e-mail: anagadi@ub.edu.sa

Preface

The role of the Internet has gradually changed a lot because of the development in various communication technologies. Now, a day's billions of people and physical devices are connected via the Internet. In the near future, storage and computational services will be more pervasive and distributed. Even today, we can see that people, machines, objects and platforms are connected with wireless/wired sensors. With the advent of Internet of Things, a deluge of devices is being connected to the network and is generating a large amount of data. According to Gartner research, 75 billion "things" will be connected to the Internet by 2025. The present IoT system relies on centralised model. The connection, authentication and data processing of the devices are managing using cloud servers that require high processing power and storage capacity even if they happen to be a few feet apart. This model is good for small-scale IoT; however, the growing needs of the large IoT environment will need a decentralised model. The decentralised IoT environment would not only reduce the infrastructure cost but provide standardised peer-to-peer communication model for the hundreds of billions of transactions. However, peer-to-peer communication has a big challenge of security. The idea of blockchain can be implanted to IoT networks to deal with the issue of scale, trustworthy and decentralisation. However, the limited processing power, storage size and energy consumption of IoT device are a major point of concern for blockchain cryptographic functions. Moreover, the efficiency, reliability and interoperability among blockchains still need to be studied and addressed.

The book presents practical as well as conceptual knowledge of the latest trends, tools, techniques and methodologies of blockchain for Internet of Things. Each chapter presents the reader with an in-depth investigation regarding the possibility of blockchain in Internet of Things. The book presents the state-of-the art and future perspectives of decentralised Internet of Things, where industry experts, researchers and academicians had shared ideas and experiences surrounding frontier technologies, breakthrough and innovative solutions and applications.

Organisation of the Book

The book is organised into three parts and altogether 11 chapters. First part, titled "Fundamentals of Decentralised IoT and Role of Blockchain", and contains chapters "Decentralised Internet of Things", "Practical Privacy Measures in Blockchains", "Empirical Evaluation of Blockchain Smart Contracts". Second part, named "Blockchain Languages, Algorithms, Frameworks and Simulation", contains chapters "Blockchain Frameworks", "Consensus Algorithm", "Smart Contracts-Enabled Simulation for Hyperconnected Logistics", "Validating BGP Update Using Blockchain-Based Infrastructure". Third part, titled "Use Cases and Applications", contains chapters "Blockchain and Smart Contract in Future Transactions—Case Studies", "IoMT: A Blockchain Perspective", "Legal Ramifications of Blockchain Technology", "On the Opportunities, Applications, and Challenges of Internet of Things". A brief description of each of the chapters is as follows:

Chapter "Decentralised Internet of Things" presents the fundamental concepts of blockchain and Internet of Things. The chapter presents discussion on the problem of growing number of IoT devices that brings challenges to the existing centralised computing system. The blockchain and investigation about the feasibility of the blockchain in Internet of Things settings are the focus of the chapter.

Chapter "Practical Privacy Measures in Blockchains" presents the discussion on privacy measures in blockchain which are practically required. The chapter discusses blockchain components such as hash, asymmetric cryptography, digital signatures, peer-to-peer network protocols and "proof of correctness/work" resulting from a game-like setup. The chapter has investigated techniques that can be used to successfully manage privacy in the blockchains. The chapter also presents an analysis on the performance evaluation of blockchains in managing privacy in Hyperledger Fabric platform.

Chapter "Empirical Evaluation of Blockchain Smart Contracts" presents a detailed discussion on evaluation of smart contracts and framework. The chapter focuses on the prominent smart contract landscape for evaluation of framework to assess smart contracts. The chapter has proposed an empirical evaluation method for prominent smart contract platforms such as Ethereum, EOS for execution speeds and transaction costs.

Chapter "Blockchain Frameworks" presents discussion on popular blockchain framework for many applications. The chapters also describe criteria for selection of blockchain framework. The enterprise support, pros and cons and transaction model have been discussed to select the framework.

Chapter "Consensus Algorithm" presents detailed discussion on blockchain consensus, proof-based consensus, voting-based, Paxos-based, RAFT and soft computing-based algorithms. The chapter also presents pros and cons of various algorithms.

Chapter "Smart Contracts-Enabled Simulation for Hyperconnected Logistics" presents discussion on combination of the Internet of Things and blockchain-based technologies that represents a real opportunity for supply chain and logistics protagonists. Through design and simulation results, chapter shows how the Ethereum blockchain and smart contracts can be used to implement a shareable and secured tracking system for hyperconnected logistics.

Chapter "Validating BGP Update Using Blockchain-Based Infrastructure" presents a blockchain-based solution to secure the border gateway routing (BGP) protocol as the existing methods use of centralised database and centralised public key infrastructure (PKI). The chapter proposes a blockchain-based technology used to create a distributed or decentralised immutable database that relies on consensus of participating autonomous system (AS), to build this blockchain.

Chapter "Blockchain and Smart Contract in Future Transactions—Case Studies" presents various case studies. This presents an interesting discussion and demonstration on use cases that utilise blockchain and smart contract in real-life applications such as real estate contracts, secure certificates and intellectual property protection.

Chapter "IoMT: A Blockchain Perspective" presents several aspects concerning the evolution and importance of the Internet of Medical Devices (IoMT) applications and blockchain technology in health care. The chapter also presents IoMT applications based on blockchain for health care and aspects concerning data security in blockchain.

Chapter "Legal Ramifications of Blockchain Technology" presents a rarely addressed area of legislation and jurisdiction. This chapter discusses broad view of the relationship between blockchain and the Law, its effects on legal activities and how the Law can be used as a tool of protection in blockchain related transactions.

Chapter "On the Opportunities, Applications, and Challenges of Internet of Things" presents a study on the best potential applications, challenges and future opportunities in the area of Internet of Things. Also, the chapter discusses the general aspects and issues of IoT and explores the implication of all these in a developing country's setting taking the case of Bangladesh.

Who and How to Read This Book?

This book has three groups of people as its potential audience, (i) undergraduate students and postgraduate students conducting research in the areas of Internet of Things and blockchain; (ii) researchers at universities and other institutions working in these fields; and (iii) practitioners in the R&D departments of Internet of Things, smart cities, blockchain and many more. This book differs from other books that have comprehensive case study and data from best practices.

The book can be used as an advanced reference for a course taught at the postgraduate level in Internet of Things and blockchain.

Bisha, Saudi Arabia

Mohammad Ayoub Khan
Mohammad Tabrez Quasim
Fahad Algarni
Abdullah Alharthi

Contents

Abbreviations

ABE	Attribute-based encryption
AI	Artificial intelligence
ARTEMIS	Automatic and Real-Time Detection and Mitigation System
AS	Autonomous system
AT	Assignment Track
ATO	Australian Taxation Office
AUSTRAC	Australian Transaction Reports and Analysis Centre
BCSSs	Block-based Cloud Storage Systems
BFT	Byzantine fault tolerance
BGP	Border Gateway Routing
BMA	Basic Mass Assignment
BPA	Basic Probability Assignment
BSA	Under the Bank Secrecy
BTRC	Bangladesh Telecommunication Regulatory Commission
CA	Consensus Algorithms
CAGR	Compound annual growth rate
CCA	Central Certificate Authorities
CCTV	Closed-circuit television
CEP	Complex event processing
CI	Cognitive intelligence
CrAN	Crowd-associated network
CSS	Cloud storage systems
DAG	Directed acyclic graph
DAOs	Decentralized autonomous systems
DApps	Distributed applications
DEX	Decentralised exchange
DLT	Distributed ledger technology
DoS	Denial-of-service
DSA	Double-spending attack
DST	Dempster–Shafer theory

EHR	Electronic health record
EHRs	Electronic health records
EOAs	Externally owned accounts
EPC	Electronic Product Code
EVM	Ethereum Virtual Machine
FinCEN	Financial Crimes Enforcement Network
FT	Fault-tolerant
GPS	Global Positioning System
HDG	Healthcare Data Gateway
HI	Human intelligence
IBAPV	Identity-based aggregate path verification
ICO	Initial Coin Offering
id2r	Identity-based inter-domain routing
IoMT	Internet of Medical Devices
IoSA	Internet of smart agriculture
IoSC	Internet of smart cities
IoSH	Internet of Smart Health
IoSI	Internet of smart industry
IoSL	Internet of smart living
IOTA	Internet of Things applications
IPC	Inter-process communication
IRIS	Intelligent Retinal Imaging System
ITAS	Innovative Technology Arrangement and Services
JVM	Java virtual machine
KSI	Keyless Signature Infrastructure
LCD	Liquid crystal display
LoRA	Long-range wireless communication
LPWAN	Low-power wide-area network
LSE	London Stock Exchange
MADM	Multi-attribute decision making
MCDM	Multi-criteria decision making
MDIA	Malta Digital Innovation Authority
MPC	Multi-party computation
NEM	New Economy Movement
NFC	Near-Field Communication
NIZK	Non-interactive zero-knowledge
NMLS	Nationwide Multistate Licensing System
PBCA	Proof-based Consensus Algorithm
PBFT	Practical Byzantine Fault Tolerance
PDAs	Personal digital assistants
PDC	Private data collection
PKI	Public key infrastructure
PoA	Proof-of-Authority
Pob	Proof-of-Burn
PoC	Proof-of-Capacity

PoET	Proof-of-Elapsed-Time
POI	Proof-of-Concept
PoI	Proof-of-Importance
PoS	Proof-of-Stake
POW	Proof-of-Work
QoS	Quality of service
R&D	Research and Development
RIR	Regional Internet registry
ROI	Return on investments
RPC	Remote Procedure Call
RPCA	Ripple Protocol Consensus Algorithm
SA	Sensitivity analysis
SBFT	Speculative Byzantine Fault Tolerance
SCM	Supply chain management
SEE	Sandboxed Execution Environment
SIM	Subscriber identity module
SPOF	Single point of failure
UX	User Experience
VBCA	Voting-based Consensus Algorithm
VNI	Visual Networking Index
WLAN	Wide Local Area Networks

Fundamentals of Decentralised IoT and Role of Blockchain

Decentralised Internet of Things

Mohammad Ayoub Khan, Fahad Algarni and Mohammad Tabrez Quasim

Abstract The growing number of IoT devices brings challenges to the existing centralised computing system. The existing security protocols are unable to protect the security and privacy of the user data. The current IoT system rely on centralised model. The decentralised IoT system would not only reduce the infrastructure cost but provide standardised peer-to-peer communication model for the massive transactions. However, peer-to-peer communication model has a big challenge of security. The blockchain technology ensures transparent interactions between different parties in a more secure and trusted way using distributed ledger and proof-of-work (POW) consensus algorithm. Blockchain enables trustless, peer-to-peer communication and has already proven its worth in the world of financial services. The idea of blockchain can be implanted to IoT system to deal with the issue of scale, trustworthy and decentralisation, thereby allowing billions of devices to share the same network without the need for additional resources. However, the limited processing power, storage size and energy consumption of IoT device is a major point of concern for blockchain cryptographic functions. Moreover, efficiency, reliability, interoperability among blockchain still need to be addressed. This chapter presents basic concepts of blockchain and investigation about the feasibility of the blockchain in Internet of Things settings.

Keywords Blockchain · Internet of Things · PoW · NONCE · IOTA · ADEPT · Decentralised · LPWAN

M. A. Khan · F. Algarni · M. T. Quasim (✉)
College of Computing and Information Technology, University of Bisha, Bisha, Saudi Arabia
e-mail: mtabrez@ub.edu.sa

M. A. Khan
e-mail: ayoub.khan@ieee.org

F. Algarni
e-mail: fahad.alqarni@ub.edu.sa

M. A. Khan et al. (eds.), *Decentralised Internet of Things*, Studies in Big Data 71,
https://doi.org/10.1007/978-3-030-38677-1_1

1 Introduction

The blockchain technology is the combination of innovative and matured business principles, game theory, cryptography, economics, computer science and engineering. The blockchain can be defined as a distributed and decentralised ledger that stores data such as transactions. The ledger is publicly shared across all the participants in the network. The storage consists of multiple blocks of data chained together akin to a physical chain as show in Fig. 1.

Definition A distributed data storage consisting of blocks which are linked in a backward fashion.

The blockchain is a trustless system where intermediary and centralised systems are not needed in contrast to the traditional transaction system such as banks, escrow services, clearing houses, registrars and many other such institutions. The mediation of intermediary institutions increases cost and time to settle a transaction, and also limits the transaction sizes. The mediation acts as arbitrator to settle disputes, however, completely non-reversible transaction was never possible with traditional systems.

The decentralised system leads to truly trustless system. Thus, eliminate the necessity for the third-party intermediaries and middleman to process the transactions. The processing of such transactions by the third-party could attract fees for the services. However, in the blockchain technology two nodes can do transaction with each other directly without having to rely on the intermediary institution such as bank, Accountant, certifying authority (CA) [3]. Since, the system is decentralised, therefore, there is no one central point of hacking or failure. The hacker needs to take over millions of nodes and computers to hijack the blockchain network which is currently not possible with the available computing power.

We have billions of IoT devices everywhere and many more are yet to join the IoT ecosystem. The IoT ecosystem has different manufacturing makes, models, and communication protocols which makes difficult to control and exchange the data. Thus, in present scenario the centralised IoT system has lots of challenges.

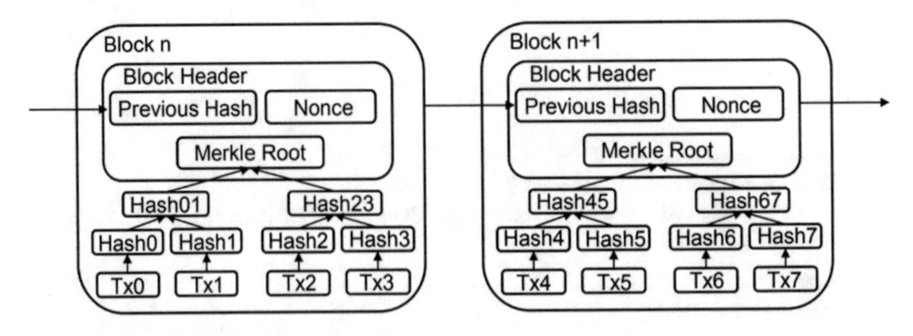

Fig. 1 General structure of blockchain [1–3]

Table 1 Characteristics of centralised and decentralised system

Characteristics	Centralised	Decentralised
Control	Central	Distributed
Design	Easy	Difficult
Maintenance	Easy	Difficult
Failures	Single point	No single point
Stability	Low	High
Vulnerable	Yes	No
Unethical operation	Possible	Not possible
Scalability	Low	High

However, the blockchain can be used to build a decentralised system which is peer-to-peer system for the IoT devices to communicate with each other.

Some of the initiative in this area is ADEPT (Autonomous Decentralised Peer-to-Peer Telemetry) from IBM and Samsung that has developed a bitcoin-based platform to build a distributed network of devices. The ADEPT uses BitTorrent for file sharing, Ethereum for smart contracts, and TeleHash for peer-to-peer messaging in the platform [4].

1.1 Centralised vs Decentralised System

A distributed system can be centralised or decentralised. In a centralised distributed system, the master node is responsible for breaking down computational tasks and data to distribute the load across the network. While as in a decentralised distributed system there is no master node, every node is autonomous to distribute the computational load across the network. In the Table 1, we have present some of the characteristics of centralised and decentralised system.

1.2 Elements of Blockchain

1.2.1 What Is a Block

The blockchain is an ordered, linked list of blocks that contains transactions. The block has an address of previous block in the chain. The first block in the blockchain is called *genesis* block. A block is not limited to transactions only but can hold any type of data such as multimedia data, software program, currency etc. [1, 3].

(A) Block Unique Identity

The block is identified by a hash value that is generated using hash function generally, SHA256. The hash value is stored in the header of the block. The header field contains hash value of *previous* block or *parent* block. The sequence of linking block to previous block creates a chain going back to the first block, known as the *genesis block.*

(B) Structure of Blockchain

The block consists of a header, metadata and a list of transactions. A block header is fixed 80 bytes, whereas size of transaction is not fixed and varies depending on the type of application (Fig. 2).

Block Header: The block header consists of metadata as shown in Table 2.

Importance of Previous Block Hash: Because the current block keeps the hash value of previous block therefore, no one can alter any transaction in the previous block. If there is any change in the transaction then it would get validated by all contributing nodes. Thus, every node in the blockchain has its own copy which is consistent across the chain. We can summaries characteristics of the blockchain transaction in Table 3.

Fig. 2 Fields of block [3]

Block Size: 4 Bytes
Block Header: 80 Bytes
Transaction Counter: 1-9 Bytes
Transaction: Variable

Table 2 Block header structure [1–3, 5]

Size	Field	Description
4 bytes	Version	A version number of software and protocols
32 bytes	Previous block hash	A reference to the hash of the previous block
32 bytes	Merkle root	A hash of the root of the merkle tree of this block's transactions
4 bytes	Timestamp	The approximate creation time of this block
4 bytes	Difficulty target	The proof-of-work algorithm difficulty target for this block
4 bytes	Nonce	A counter used for the proof-of-work algorithm

Table 3 Characteristics of blockchain [3]

Communication type	P2P
Transaction model	Shared and open
Database type	Append only
Third-party authentication server	No
Transaction nature	Immutable

1.2.2 What Is a Miner?

The blockchain (ledger/storage) is stored in the computers of many people. These computers or people are called nodes. The node can be a simple users or miners. However, miner can be terms as a special node in the blockchain network that gives its computational power to the network.

Every node in the blockchain has a copy of all transactions ever made. Any new transactions in the chain triggers the update of blockchain. When a new transaction triggers then miners add a new block of transactions to the blockchain, after due verification as shown in Fig. 3. As a reward for their efforts, miners may be given cryptocurrency whenever they add a new block of transactions to blockchain.

(A) Consensus

Consensus means come to a general agreement among entities. Since the blockchain is open for appending information to the block, hence, a malicious node may attempt to add false information into block. Therefore, consensus method is adopted by blockchain technology to invalidate such malicious attempts. Essentially, a consensus method is employed to allow all the nodes in the chain to come to agreement for

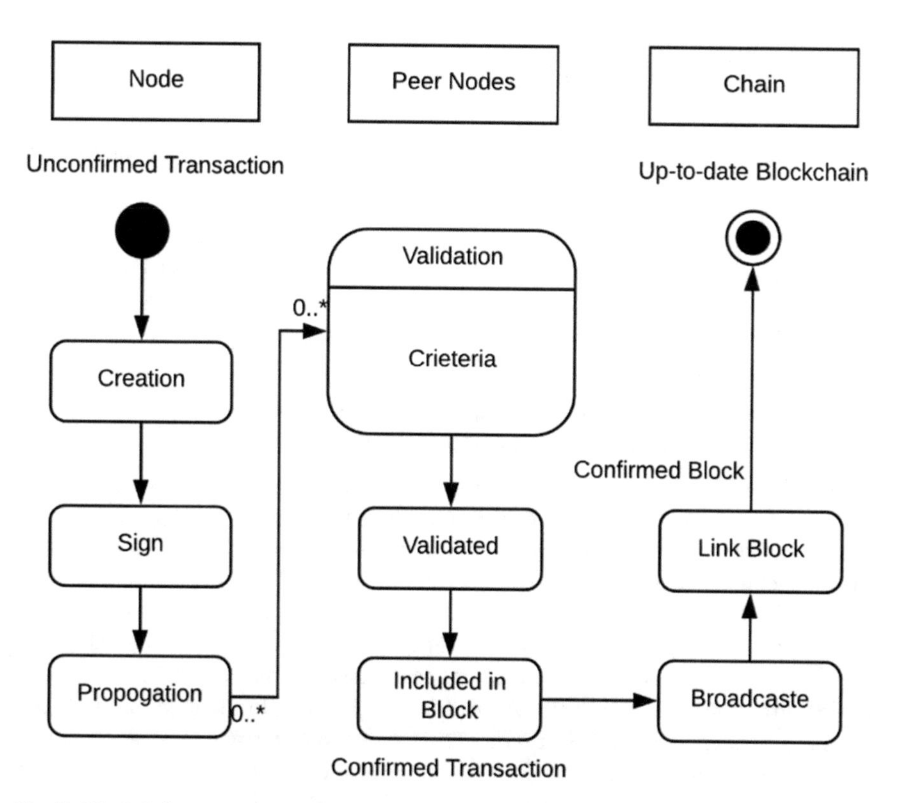

Fig. 3 Blockchain transaction cycle

true information. This is also important to mention here that if a malicious node can manage to achieve majority control of the network then the consensus will be bypassed and malicious node will have full control on the chain. There are many methods based on computational power, system stake and inter-network relationship to achieve consensus. We have listed some of the consensus methods as follows:

- Proof-of-Work
- Proof-of-Stake
- Proof-of-Activity
- Proof-of-Importance
- Proof-of-Capacity
- Proof-of-Burn
- Proof-of-Weight
- Delegated Proof-of-Stake
- Leased Proof-of-Stake
- Proof of Elapsed Time
- Practical Byzantine Fault Tolerance
- Simplified Byzantine Fault Tolerance
- Delegated Byzantine Fault Tolerance
- Directed Acyclic Graphs

To understand the underlying mechanism of blockchain, we will only discuss Poof-of-Work in this chapter. The Proof-of-Work is the first and straight forward algorithms introduced in the blockchain network. Most of the blockchain technologies uses this model to confirm all of their transactions and produce relevant blocks to the network chain.

(B) Proof-of-Work (PoW)

A process, where miners solves a complex computational puzzle in order to get privilege to add a block of transactions to the blockchain. The underlying principle behind miners is to solve complex mathematical problems and easily give out solutions for verification. Therefore, the objective of PoW is to find or generate a value which is:

- Difficult to generate as it requires high computational power
- But, easy to verify by even small computational power.

NONCE—The difficulty level is mentioned in *nonce* field of the header. A number is added to a hashed block so that when rehashed, meets the difficulty level restrictions. The miner needs to search this number before solving puzzle for a block in the blockchain. Then finds all possible hash that will make you to reach the target. This random number can be anything between 0 and 2^{32}. The process of determining *nonce* is called *mining*. The *nonce* requires significant amount of trial-and-error, heuristic, as it is a random string. A miner shall guess a string, append it to the hash of the current header, rehash the value, and compare this to the target hash. If the resulting hash value meets the criteria of *nonce* then block is win. The block is then added to the chain. Also, *nonce* is used to check the validity of the hash. To find a valid hash, we need to find a *nonce* value that will produce a valid hash when combined with the remaining information of the block.

2 Security Issues in Internet of Things

The IoT is an ecosystem of heterogeneous devices with embedded sensors interconnected through a network, as shown in Fig. 4. The IoT devices are uniquely identifiable using Electronic Product Code (EPC), IPv6, and URI etc.

The IoT device has low power, tiny memory and limited processing power. A hardware called gateways can be deployed to connect IoT devices to the outside world for data exchange. The communication protocol is still not matured and yet to be standardise. Different vendors use different protocol for identification, communications, discovery and device management as shown in Table 4.

2.1 Security Requirements for IoT

The conventional cryptographic algorithms are not sufficient and appropriate to provide security infrastructure in a heterogenous environment of IoT [11]. Since, IoT is based on internet, therefore, is inherently insecure as the data security has been designed in ad hoc manner as when required [12]. Furthermore, IoT is a resource constraint which has limited computing capability as it continuously generates, exchanges and consumes data with minimal human interventions [13]. With that

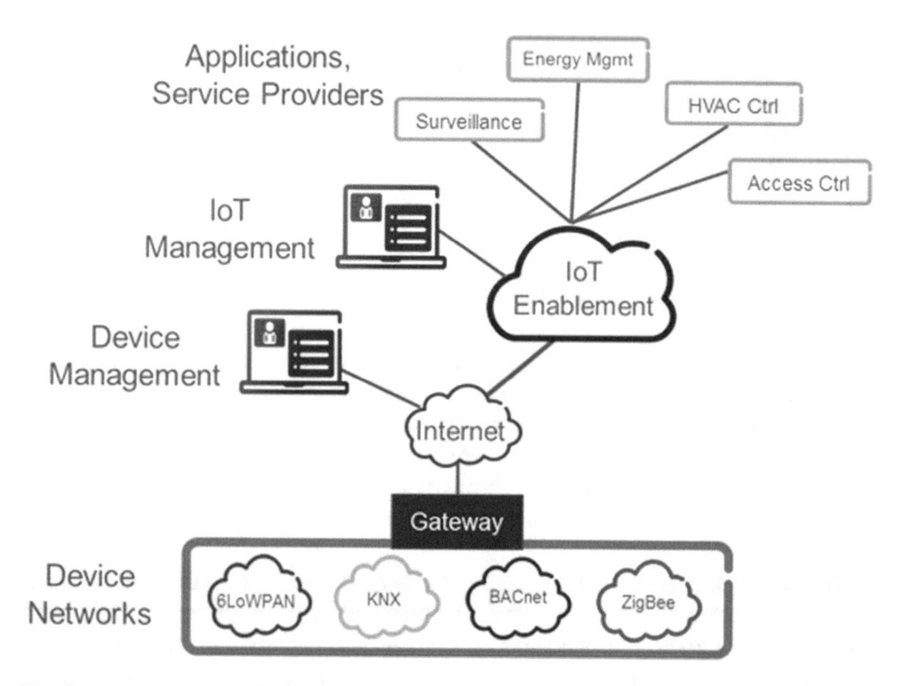

Fig. 4 Internet of Things [6]

Table 4 Protocols in IoT [7–10]

Protocols	Example
Identification	EPC, uCode, IPv6, URIs
Network	6LowPAN, IPv4/IPv6, RPL, CORPL, CARP
Transport	Wifi, Bluetooth, LPWAN, 4G, LTE, NFC
Discovery	Physical Web, mDNS, DNS-SD
Data protocols	MQTT, SMQTT, CoRE, CoAP, AMQP, Websocket, Node
Device management	TR-069, OMA-DM
Semantic	JSON-LD, Web thing model
Multi-layer frameworks	Alljoyn, IoTivity, Weave, Homekit
Authentication	Oauth2, OpenID, OMA DM, LWM2M, TR-069, PKI

said, it is not feasible to extend computationally extensive and costly internet security solutions to IoT [14].

The data in the IoT system is stored and accessed using cloud services [15–17] which are prone to cyber-attacks such as SQL injection, tampering, and single-node failure [18–20]. In general, the requirement of security in IoT is depicted in Fig. 5.

Being distributed and decentralised, the blockchain has potential to address the security issues of IoT that includes data integrity, reliability, and authentication [1]. The data captured from sensor events such as moisture, humidity, temperature, location, blood pressure can be stored in tamper resistant ledgers that are readable only to authorised nodes in the network.

The following prominent features of blockchain can contribute to the integrity of IoT applications and so enhance the IoT security.

2.1.1 Data Privacy, Confidentiality and Integrity

The information leakage in IoT system is obvious as it connected to internet. The data may become accessible to various organisations and domains across the internet. The privacy becomes important as the data is likely to be exposed to sophisticated malicious parties and therefore increases the probability of being exploited and attacked. Generally, IoT data passes through multiple hops in a network, therefore, a lightweight encryption mechanism is needed to ensure the privacy and confidentiality of the data.

This is also difficult for IoT system to provide confidentiality for data transmission due to the resource constrained nature of low-end devices. Therefore, a light-weight cryptographic algorithm is needed. The IoT devices are placed in unattended environment that makes data integrity a concern. Tampering data is much easier task in unattended wireless network. Further, IoT devices is quite likely to have low quality

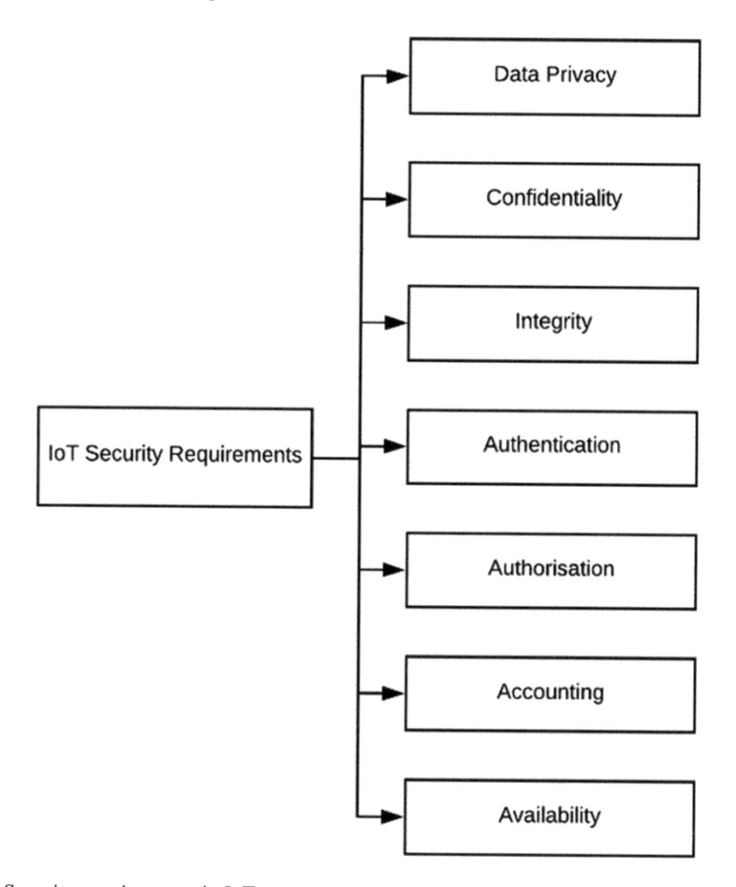

Fig. 5 Security requirements in IoT

or corrupted data therefore, we also need the integrity algorithm which is best suited to IoT system. The integrity algorithm like SHA and MD5 can be deployed.

2.1.2 Authentication, Authorisation and Accounting

The IoT devices such RFID tags and wireless sensors are deployed in unattended public areas without any protection which makes it vulnerable to physical attacks. An illegitimate sensor can do location impersonation by registering and claiming that it is at one location while it is actually at a different location. Therefore, device authentication become extremely important that involves recognizing the device and verifying its association with a correct topological address.

In order to exchange the data from one device to another device in IoT, the devices must be authenticated. However, due to diverse heterogeneous underlying architectures this becomes challenging to define standard global protocol for authentication and authorisation in IoT. Also, accounting and audit trail of resources is crucial area to address in IoT system.

2.1.3 Availability of Services

The attacks on IoT devices may hinder the provision of services through the conventional Denial-of-Service (DoS) attacks. Various strategies including the sinkhole attacks, jamming adversaries or the replay attacks exploit IoT components at different layers to deteriorate the Quality-of-service (QoS) being provided to IoT users.

2.1.4 Energy Efficiency

The IoT devices are resource constrained and characterised with low power and less storage. The attacks on IoT architectures may result an increase in energy consumption by flooding the network and exhausting IoT resources.

2.1.5 Single Point of Failure (SPOF)

The growing number of IoT devices in the heterogeneous network may expose a large number of single-point-of-failure as the network architecture is centralised. Therefore, it necessitates the development of a tamper-proof and fault tolerant environment for a large number of IoT devices.

2.2 Security Analysis

Traditional security algorithms such as the asymmetric encryption, are computationally extensive for resource constraint IoT devices. The data received from sensors can be stored, forwarded and processed by many different intermediate systems which are prone to be tampered and forged. There are number of attacks as listed below which can happen in IoT system.

- Attacks to end devices
- Attacks to communication channels
- Attacks to network protocols
- Attacks to sensory data
- Denial of service (DoS) attack
- Software attacks

3 Feasibility of Blockchain Technology

The sensory data can be protected by digital signatures in a transactional fashion that guarantees that the data is sent by the only authorised IoT devices are recorded and exploited. The backward-linked hashed structure enhances the trustworthiness of sensory data recorded in IoT-Blockchain ledgers. There is a blockchain based IoT platforms developed by Samsung Electronics, IBM [4], and advocated device democracy to be the future of IoT known as Autonomous Decentralised Peer-to-Peer Telemetry (ADEPT) [2, 4, 21, 22].

3.1 Transaction of Sensory Data

The blockchain can be used to consider the raw sensory data as transaction, verifying transactions, and even mining blocks. The IoT participating nodes can be configured to define as ordinary node or miners in the network. The miner nodes in IoT, mines transactions into block and store all blocks that requires large storage and high computational resources. The ordinary node stores all the information in the blocks that includes block headers and block data, however, this node doesn't participate in block mining process. The IoT devices can generate private keys independently, or register with the certificate authority (CA) for access control and audit. The new devices need to register and enroll in the CA first to maintain the private keys. The private keys of the nodes can be applied to generate signature of transaction to claim the ownership of transaction. The IoT node generate and broadcast transactions in order to be added in the block.

Here, private keys possessed by clients (light nodes) are applied to generate signatures of transactions to valid the owners of transactions. The clients only generate and broadcast transactions. The sensors have only sensing capability therefore, sensors are unable participate in blockchain functions. The local acquisition node can interpret the collected sensory data as transactions and broadcast the transactions into the blockchain network [5].

The existing architectures and protocols of blockchain are designed which is best suited to powerful peer-to-peer computing homogeneous network. The IoT is a limited resource of end devices as compared to high-performance servers or desktop computing devices, prevent directly deploying blockchain in IoT. There are many challenges in applying blockchain in IoT as discussed in the next section.

3.2 Computational Speed

The light-weight blockchain cryptographic algorithms like zero knowledge and Attribute-based encryption (ABE) are too heavy for IoT devices [23]. Also, the

Table 5 Hash computation speed

Processor	Speed
GPU [24]	10^7 Hashes/second
Raspberry Pi [25]	10^4 Hashes/second
Bitcoin Network [3]	10^{19} Hashes/second

Table 6 Hash computation speed

Framework	Blocks	Space
Bitcoin [3]	5×10^5	150 GB
Ethereum [24]	5×10^6	400 GB

consensus protocols are computationally hard to run on IoT devices. The speed of hash computation for different processors are shown in Table 5. Therefore, this is obvious that IoT devices with large number of nodes are unable to contribute enough computational resources that can perform PoW computations.

3.3 Storage Growth

With the growth of transaction in blockchain, the number of blocks also increases that demands for large storage space. The number of blocks and space of Bitcoin/Ethereum is shown in Table 6.

In order to verify the transactions generated by others the IoT must store all this massive data which is a constraint when there are large number of IoT devices in the network. Therefore, mass storage can be attached with IoT device which is a simple and straight forward solution. On the other hand, the IoT device can trust remote servers for storing blockchain, however, this will be an extra overhead to securely communicate between IoT device and remote server. Moreover, it is expensive to store data on blockchain. The price would be expensive in IoT because every block would be duplicated n times in an n-node blockchain network.

3.4 Communication Overhead

The nodes in blockchain frequently communicate in order to exchange the data. The IoT devices have shadowing, fading, and interference effect in wireless setting. There are various communication technologies that can be used in IoT as shown in Table 7.

The low-power wide-area network (LPWAN) technologies like LoRa and Sigfox which operate in unlicensed band, and cellular-based Narrow Band IoT (NB-IoT) that operates in licensed band will be a game changer for deployment of IoT devices [7]. We have shown features of LPWAN communication technologies in Table 8 [10].

Table 7 Hash computation speed [26]

Technology	Speed	Energy consumption
IEEE 802.15.1 (Bluetooth)	720 kbps	140 mJ/Mb
IEEE 802.15.4 Zigbee	250 kbps	300 mJ/Mb
IEEE 802.15.3 (UWB)	110 Mbps	7 mJ/Mb
IEEE 802.11 a/b/g (Wi-Fi)	54 Mbps	13 mJ/Mb

Table 8 LPWan IoT connectivity [9, 10]

Features	LoRa	Sigfox	NB-IoT	eMTC	EC-GSM
Range	<11 km	<13 km	<15 km	<11 km	<15 km
Bandwidth	<500 kHz	100 Hz	180 kHz	1.08 MHz	200 kHz
Data rate	<50 Kbps	<100 bps	<170 Kbps	<1 Mbps	<140 Kbps
Frequency spectrum	Unlicensed	Unlicensed	Licensed LTE	Licensed LTE	Licensed GSM

The short-range communication technologies like NFC, Bluetooth, ZigBee limits the coverage area, transmission rate and topology for IoT devices [8].

3.5 Energy Consumption and Latency

The IoT devices are inherently designed to operate for a long time with battery energy source. The average power consumption in IoT device is 0.3 mWh/day that lasts for 5 years using a CR2032 battery with the capacity of 600 mWh [27]. The energy consumption in normal communication using different technologies is shown in Table 7. On average, a SHA-256 computation requires around 90 nJ/B [28]. If we assume the transmission in IoT using IEEE 802.15.1, then the energy budget of 0.3 mWh per day can only support about 0.5 MB data processing which is very small for blockchain. The IoT is a delay-sensitive application as many decisions depends on sensor data. However, in blockchain the delay is tolerable.

3.6 Sensory Data Privacy Issues

The sensitive data generated from the sensors should be kept confidential. There are many encryption algorithms like Identity Based (IBE), Attributed-based Encryption (ABE) that is applied. Thus, the data in transactions can be encrypted and miners or users can decrypt using decryption credential.

3.7 Device Identity and Management

In public blockchain, peers are defined by the public addresses that can be created independently without prior notification to others. While as in private blockchains, peers need to be authorised to enter in the blockchain network. Therefore, the identity management becomes basic requirement of private blockchains. In IoT applications, the owners should know the identities of the devices and devices must know their owner [29]. The IPV6 can be used to identify the device uniquely across the network.

4 Blockchain Technology in IoT

In this section, we will discuss typical technologies of blockchain which can be used in IoT applications. Based on access control of the blockchain network, the state-of-the-art blockchain can be categorised into public, private, and hybrid blockchains as shown in Table 9.

The public blockchain is suitable for IoT due to its open accessibility and flexibility. On the other hand, private blockchain is a proprietary network with stringent access control. The read/write permission, as well as participant identification and certification are restricted in the private blockchain [30, 31]. However, due to low computational complexity the private blockchain are faster than public blockchain. A simplified protocol called Practical Byzantine Fault Tolerance (pBFT) is used to restrict access control. The access control provided by private blockchain further protects IoT application from external attacks [32]. Therefore, private blockchain is best suited for IoT applications with small number of miners and nodes. The hybrid blockchain exploits the block generation rate of private blockchains, and scalability feature of public blockchains [33]. The hybrid blockchain is attractive for IoT applications due to the complexity and heterogeneity of IoT system.

All the three blockchain categories can be applied to different applications depending on the complexity and heterogeneity of IoT system. The consensus protocols play an important role to determine the performance of blockchain based IoT applications. The PoW based consensus protocols are found to be most secure in open networks [34]. Among many blockchain frameworks, Ethereum is found to be appropriate for many IoT applications with large numbers of IoT devices and heterogenous environment.

Table 9 Comparison of public, private and hybrid blockchains

Type	Access	Read	Participation	Contribution	Scalability	Comp. complexity	Example
Public	Yes	Yes	Yes	Yes	High	High	Bitcoin, Ethereum
Private	Stringent	Stringent	Stringent	Stringent	Low	Low	Hyperledger
Hybrid	Mixed	Mixed	Mixed	Mixed	Medium	Medium	NA

4.1 IOTA—A Decentralised Approach

The word "iota," means "a small amount," to be used in micropayments-based system. The IOTA has completed all the preparations for decentralisation [35, 36]. The IOTA is designed to address the limitations of blockchain. The IOTA is a decentralised cryptocurrency platform for Internet of Things applications. The distinguish part of the IOTA is the Tangle, that is built for machine-to-machine (M2M) communication [37]. Also, IOTA is ready to solve the scalability issues facing by modern Internet. The IOTA has focus on M2M communication that can support many applications like smart home, smart city technology, smart healthcare.

The DAG data structure of Tangle is capable of handling the IoT-driven large volume of data traffic because DAG allows transactions to be issued simultaneously, asynchronously, and continuously [35].

The IOTA Foundation has already formed partnerships in some key IoT verticals, such as with Jaguar Land Rover [38]. Many companies like IoTex, Oracle are putting efforts to combine connected devices with the Hyperledger-based blockchain solutions. However, still there is no matured product available in the market.

The IoT can enhance many dimensions of daily lives by technology integrated with invaluable data. For example, energy saving, smart thermostats, smart transportation, smart home, Smart alarms and surveillance systems are to name a few. On the other side of the benefits, there are of course worrying downsides as well. The applications devices like Amazon's Alexa, mobile phones, internet browsers, social networking apps are collecting billions of pieces of information every day that includes browser habits, location information, your biometrics, facial ID and many more. This data is often distributed largely to other parties beyond our control and reach.

4.2 Proposed System Architecture

The blockchain and IoT can be integrated as shown in Fig. 6. The data is received by the data logger from many sensors. The data logger sends the encrypted data to the local acquisition node which is computationally powerful. The local acquisition node is integrated to a LPWAN gateway. The peers in this network sends data through the gateway, the data is stored in the blockchain nodes. A brief discussion on various components of IoT-Blockchain is presented in the next section.

4.2.1 LPWAN Gateway (Full Blockchain Node)

The gateway is a powerful computing device that acts as full node in the blockchain. The primary responsibility is to route the data to the network and verify integrity to achieve a trustless IoT infrastructure.

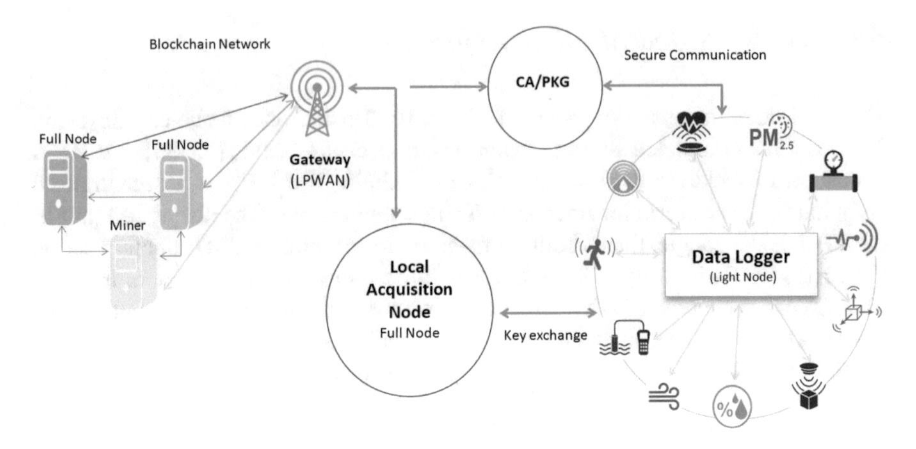

Fig. 6 Proposed IoT-blockchain platform system architecture

4.2.2 Local Acquisition Node (Full Blockchain Node)

The primary responsibility of the local acquisition is to acquire the data sent by the data logger. The data received from the logger is encrypted to ensure the confidentiality. The key distribution is achieved with the help of certifying authority or key generators. Local acquisition node sends the transaction to LPWAN which is a full node to solve the puzzle or PoW.

4.2.3 End Devices (Sensors)

This is true that blockchain functions are not possible to implement in sensors as they have only sensing capabilities. The data which is received by the IoT /LPWAN gateway are pushed to a blockchain infrastructure as shown in Fig. 6. This is one way to communicate with a blockchain node without any storage or computational requirements in IoT system.

5 Conclusion

Internet of Things is a key enabler for many contemporary digital transformation domains such as smart cities, transportation systems, energy systems, healthcare and others. The state-of-the-art blockchain technologies were discussed in this chapter followed by comparison of technologies in terms of protocol, speed, and applicability to the IoT scenarios. We have also proposed an architecture for integration of blockchain and Internet of Things with LPWAN gateways and data acquisition node.

References

1. Zheng, Z., Xie, S., Dai, H.-N., Chen, X., Wang, H.: Blockchain challenges and opportunities: a survey. Int. J. Web Grid Serv. **14**, 352 (2018). https://doi.org/10.1504/ijwgs.2018.095647
2. O'Connor, C.: What Blockchain means for you, and the Internet of Things (2017). https://www.ibm.com/blogs/internet-of-things/watson-iot-blockchain/. Accessed 20 Sept 2019
3. Blockchain (2017). https://blockchain.info. Accessed 15 Sept 2019
4. Panikkar, B., Nair, S., Brody, P., Pureswaran, V.: ADEPT: an IoT practitioner perspective. IBM (2014)
5. Dorri, A., Kanhere, S.S., Jurdak R.: Towards an optimized blockchain for IoT. In: Proceedings 2nd International Conference on Internet-of-Things Design and Implementation, pp. 173–178. ACM (2017)
6. Höller, J.: Having a headache using legacy IoT devices? https://www.ericsson.com/en/blog/2012/11/having-a-headache-using-legacy-iot-devices. Accessed 30 Sept 2019
7. Ericsson, Cellular networks for massive IoT. https://www.ericsson.com/res/docs/whitepapers/wpiot.pd. Accessed 1 Sept 2019
8. Vangelista, L., Zanella, A., Zorzi, M.: Long-range IoT technologies: the dawn of LoRa. In: Future Access Enablers of Ubiquitous and Intelligent Infrastructures, pp. 51–58. Springer (2015)
9. Bardyn, J., Melly, T., Seller, O., Sornin, N.: IoT: The era of LPWAN is starting now. In: Proceedings of the 42nd European Solid-State Circuits Conference, ESSCIRC Conference, pp. 25–30. Lausanne, Switzerland, IEEE (2016)
10. Nokia, LTE evolution for IoT connectivity (2015). http://resources.alcatel-lucent.com/asset/200178. Accessed 21 Sep 2019
11. Katagi, M., Moriai, S.: Lightweight Cryptography for the Internet of Things (2012)
12. Fabian, B., Günther, O.: Security challenges of the EPC global network. Commun. ACM. **52**(7), 121–125 (2009)
13. Rose, K., Eldridge, S., Chapin, L.: The internet of things: an overview. In: The Internet Society, pp. 1–50 (2015)
14. Weber, R.H.: Internet of Things—new security and privacy challenges. Comput. Law Secur. Rev. **26**(1), 23–30 (2010). https://doi.org/10.1016/j.clsr.2009.11.008
15. Liu, Y., Dong, B., Guo, B., Yang, J., Peng, W.: Combination of cloud computing and Internet of Things (IOT) in medical monitoring systems. Int. J. Hybrid Inform. Technol. **8**(12), 367–376 (2015)
16. Atlam, H.F., Alenezi, A., Alharthi, A., Walters, R.J., Wills, G.B.: Integration of cloud computing with Internet of Things: challenges and open issues. In: Proceedings IEEE International Conference Internet of Things (iThings) and IEEE Green Computing and Communications (GreenCom) and IEEE Cyber, Physical Social Comput. (CPSCom) and IEEE Smart Data (SmartData), pp. 670–675 (2017). http://dx.doi.org/10.1109/iThings-GreenComCPSCom-SmartData.2017.105
17. Lyu, X., Ni, W., Tian, H., Liu, R.P., Wang, X., Giannakis, G.B., Paulraj, A.: Optimal schedule of mobile edge computing for internet of things using partial information. IEEE J. Sel. Areas Commun. **35**(11), 2606–2615 (2017)
18. Booth, G., Soknacki, A., Somayaji, A.: Cloud security: attacks and current defenses. In: Proceedings 8th Annual Symposium on Information Assurance, ASIA13, pp. 4–5. Citeseer (2013)
19. Chidambaram, N., Raj, P., Thenmozhi, K., Amirtharajan, R.: Enhancing the security of customer data in cloud environments using a novel digital fingerprinting technique. Int. J. Digit. Multimed. Broadcast. **2016**, 1 (2016)
20. Kshetri, N.: Can blockchain strengthen the Internet of Things? IT Prof. **19**(4), 68–72 (2017)
21. Brody, P., Pureswaran, V.: Device democracy: saving the future of the internet of things. IBM (2014). https://public.dhe.ibm.com/common/ssi/ecm/gb/en/gbe03620usen/global-business-services-global-business-services-gb-executivebrief-gbe03620usen-20171002.pdf. Accessed 2 Sept 2019

22. IBM: Watson Internet of Things (2017). https://www.ibm.com/internet-ofthings/. Accessed 2 Sept 2019
23. Rahulamathavan, Y., Phan, R.C.W., Misra, S., Rajarajan, M.: Privacy-preserving blockchain based IoT ecosystem using attribute-based encryption. In: Proceedings IEEE International Conference on Advanced Network Telecommunication System. Odisha, India (2017)
24. Ethereum mining hardware (2017). https://www.buybitcoinworldwide.com/ethereum/mining-hardware/. Accessed 5 Sept 2019
25. Raspberry pi. https://www.raspberrypi.org. Accessed 2 Sept 2019
26. Lee, J.S., Su, Y.W., Shen, C.C.: A comparative study of wireless protocols: Bluetooth, UWB, ZigBee, and Wi-Fi. In: Proceedings of the 33rd Annual Conference of the IEEE Industrial Electronics Society (IECON '07), pp. 46–51 (2007)
27. Lauridsen, M., Kovacs, I.Z., Mogensen, P., Sorensen, M., Holst, S.: Coverage and capacity analysis of LTE-M and NB-IoT in a rural area. In: Proceedings 84th IEEE Vehicular Technology Conference (VTC-Fall '16), pp. 1–5 (2016)
28. Westermann, B., Gligoroski, D., Knapskog, S.: Comparison of the power consumption of the 2nd round SHA-3 candidates. In: Gusev, M., Mitrevski, P. (eds.) Proceedings 2nd International Conference on ICT Innovations, pp. 102–113. Berlin, Heidelberg (2010)
29. Roman, R., Najera, P., Lopez, J.: Securing the Internet of Things. Computer **44**(9), 51–58 (2011)
30. Yu, G., Wang, X., Zha, X., Zhang, J.A., Liu, R.P.: An optimized round-robin scheduling of speakers for peers-to-peers-based byzantine faulty tolerance. In: 2018 Proceedings IEEE Globecom Workshops (GC Wkshps '18) (2018)
31. Kravitz, D.W., Cooper, J.: Securing user identity and transactions symbiotically: IoT meets Blockchain. In: Proceedings Global Internet Things Summit (GIoTS '17), pp. 1–6 (2017)
32. Sharma, P.K., Singh, S., Jeong, Y.S., Park, J.H.: DistBlockNet: a distributed blockchains-based secure SDN architecture for IoT networks. IEEE Commun. Mag. **55**(9), 78–85 (2017)
33. Vukolić, M.: The quest for scalable blockchain fabric: Proof-of-work vs. BFT replication. In: International Workshop on Open Problems in Network Security, pp. 112–125. Springer (2015)
34. Kabessa, N., PoW vs. PoS (2017). https://medium.com/blockchain-atcolumbia/pow-vs-pos-tech-talk-77f9a1bf05d7
35. Andriopoulou, F., Orphanoudakis, T., Dagiuklas, T.: IoTA: IoT automated SIP-based emergency call triggering system for general eHealth purposes. In: 2017 IEEE 13th International Conference on Wireless and Mobile Computing, Networking and Communications (WiMob), pp. 362–369. Rome (2017). https://doi.org/10.1109/wimob.2017.8115830
36. Lamtzidis, O., Gialelis. J.: An IOTA based distributed sensor node system. In: 2018 IEEE Globecom Workshops (GC Wkshps), pp. 1–6. Abu Dhabi, United Arab Emirates (2018). https://doi.org/10.1109/glocomw.2018.8644153
37. Dasalukunte, D., Mehmood, S., Öwall, V.: Complexity analysis of IOTA filter architectures in faster-than-Nyquist multicarrier systems. In: 2011 NORCHIP, pp. 1–4. Lund (2011). https://doi.org/10.1109/norchp.2011.6126704
38. Baker, P.: Investors Pounce on IOTA as Jaguar Land Rover Announces Crypto Integration (2019). https://cryptobriefing.com/iota-jaguar-land-rover-crypto/. Accessed 20 Sept 2019

Practical Privacy Measures in Blockchains

Omar S. Saleh, Osman Ghazali and Norbik Bashah Idris

Abstract The Blockchain Technology has recently been a hot topic proposed in many industries such as Financial, Healthcare, Business, E-Government, Education, etc. The Blockchain can simply be defined as a distributed database or public ledger that contains records of all digital transactions/events that have transpired amongst the parties involved. The technology itself is comprised of other more fundamental knowledge namely: cryptography, distributed system, network and game theory. Thus at the more basic level, the blockchain components include functions such as hash, asymmetric cryptography, digital signatures, peer-to-peer network protocols and some elements of a "proof of correctness/work" resulting from a game-like setup. Against a backdrop of such a mixture of functions, "privacy" has emerged to be one of the new challenges in any Blockchain implementation. This research aims to investigate the techniques that can be used to successfully manage privacy in the blockchains. The work has identified the requirements and analyzed the techniques that can be used. Finally, the work was also extended to an analysis on the performance evaluation of blockchains in managing privacy albeit focusing on a specific blockchain—the Hyperledger fabric platform.

Keywords DLT (Distributed Ledger Technology) · Blockchain · Cryptography · Hash · Privacy · Zero-knowledge proofs · Peer-to-peer · Hyperledger

---

O. S. Saleh (✉)
Studies, Planning and Follow-Up Directorate, Ministry of Higher Education and Scientific Research, Baghdad, Iraq
e-mail: omar_saad@ahsgs.uum.edu.my

O. S. Saleh · O. Ghazali
School of Computing, University Utara Malaysia, Kedah, Malaysia
e-mail: osman@uum.edu.my

N. B. Idris
Kulliyyah of Information and Communication Technology, International Islamic University Malaysia, Kuala Lumpur, Malaysia
e-mail: norbik@iium.edu.my

M. A. Khan et al. (eds.), *Decentralised Internet of Things*, Studies in Big Data 71,
https://doi.org/10.1007/978-3-030-38677-1_2

1 Introduction

A Blockchain has been defined by [1] as a distributed database or public ledger that contains records of all digital transactions/events that have transpired amongst the parties involved. Each transaction is verified through agreement amongst the majority of parties in the system. Once a record is made, the information cannot be erased. Thus, the blockchain serves as an irrefutable record of every single transaction that has been made, thereby allowing the participating parties to know for sure that a digital event has occurred. Each transaction is contained in a block with several blocks being linked to each other linearly and chronologically in the form of a chain.

Authors [2] define the blockchain as continuous and unchangeable chains of data whereby different transactions get stored in the form of timestamped blocks. Blockchain is also known as distributed ledger technology [3]. Each copy of the Blockchain software (node) is able to store the complete copy of the ledger, write new entities to its ledger upon a given consensus among the connected nodes, broadcast transactions and regularly check its copy of ledger if its identical to the ledgers across most connected nodes [1]. The ledger is a combination of connected blocks. The block contains the transactions that took place and is chained with the previous block forming a chain hence the name blockchain. The transactions are compressed and anchored in the block using a Merkle tree model. The header of each block in the chain includes the hash of its content and the hash of all information in the previous block. In the blockchain, cryptography technology takes a significant place. It ensures the confidentiality of user data and transactions to ensure data consistency and provide all possible security. Privacy is a challenge in the blockchain due to the fact of public nature of the network. The transactions of blockchain are public; hence, it is possible to trace and extract the physical identities of the users by data mining. Privacy threats would arise from the transactions and network environment. Hence, this research describes the infrastructure of the blockchain, characteristics of the blockchain, design principles of blockchain, and the working process of the blockchain. This work also analyzes the privacy problems that blockchain still has and introduces the existing measures to these problems.

2 Blockchain Architecture

Blockchain consists of five layers, and each layer involves specific components. These layers are data layer, consensus layer, contract layer, network layer, and application layer.

1. Data layer involves several components such as data block, chain structure, timestamp, hash function, Merkle tree, and digital signature.
2. Consensus layer involves the consensus mechanisms which help the nodes to reach consensus [4].

3. The Contract layer mainly includes the smart contract and other codes which are the control logic of the decentralized application. The smart contract is a computer code embedded into the blockchain, and it comprises a set of rules [5].
4. Network layer involves the data transmission protocols and verification mechanisms [6]. A flat topology is the way of nodes connected in the blockchain, which means there is no trusted node or central node.
5. The application layer involves the applications. The typical applications of blockchain are Bitcoin, Ethereum, and Hyperledger [7]. Figure 1 shows the architecture of the blockchain and Fig. 2 shows the architecture of the blockchain in Bitcoin, Ethereum, and Hyperledger.

Bitcoin, Ethereum, and Hyperledger are the most three dominant blockchains. Many commonalities are in the overall architecture, but they are different in the

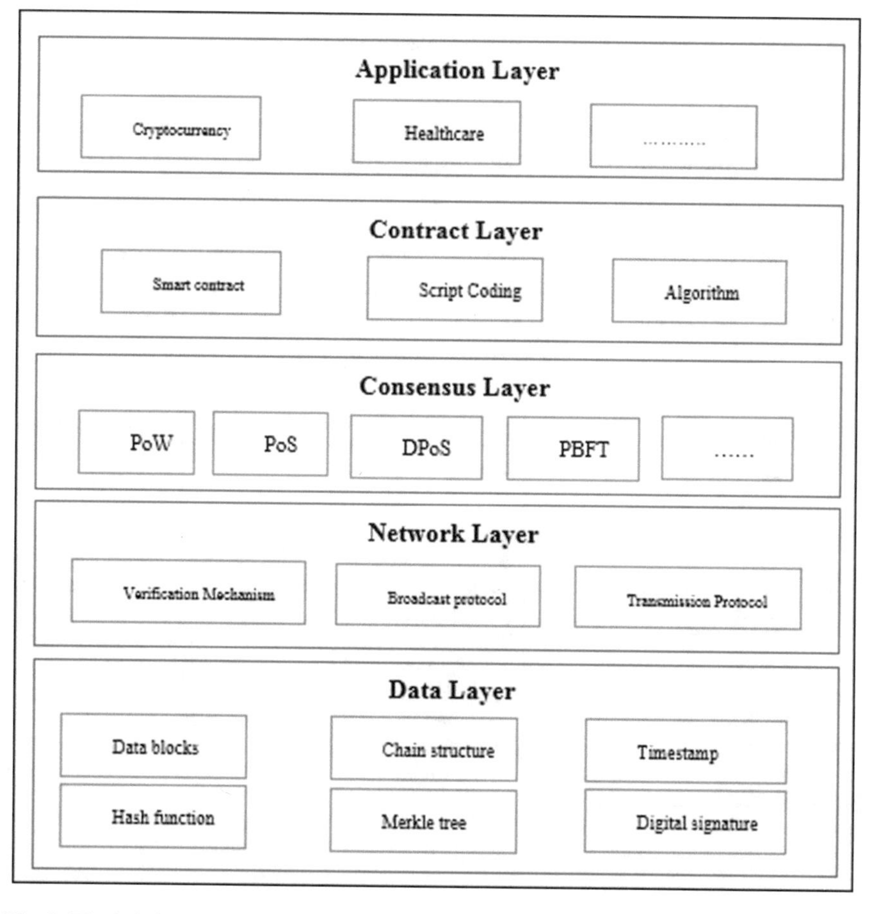

Fig. 1 Blockchain architecture

Layers	Bitcoin	Ethereum	Hyperledger
Application	Bitcoin Trading	Ethereum trading	Enterprise Applications
Contract	Script	Solidity/Script EVM	Go/Java Docket
Consensus	Pow	PoW	PBFT/SBFT
Network	TCP-based P2P	TCP-based P2P	HTTP/2-based P2P
Data	Merkle tree	Merkle Patricia Tree	Merkel Bockt tree

Fig. 2 Blockchain architecture among Bitcoin, Ethereum and Hyperledger

implementation, including data structure, consensus mechanism, smart contract, network, and application. The data structure adopted in Bitcoin is Merkle tree and in Ethereum is Merkle Patricia Tree, and Hyperledger is Merkle bockt tree. Proof of Work (PoW) is the consensus mechanism used in Bitcoin. Proof of Work (PoW) and proof of stake (PoS) are consensus mechanisms used in Ethereum. Practical Byzantine fault tolerance (PBFT) and Speculative Byzantine fault tolerance (SBFT) are consensus mechanisms used in Hyperledger [8–10]. A TCP protocol is used in both Bitcoin and Ethereum while HTTP/2 protocol is used in Hyperledger (Fig. 3).

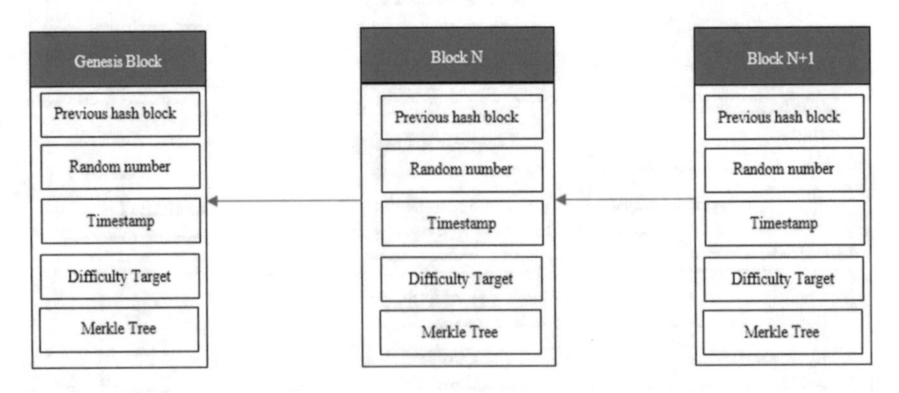

Fig. 3 Blockchain structure

2.1 Blockchain Structure

Blockchain is a chain of connected blocks. Each block is the collection of valid transactions. Any node in the blockchain-based system can start a transaction and broadcast to all nodes present in the network. Nodes in the network validate the transaction using the old transactions and then block added to the existing blockchain. Each block in the blockchain has two components which are block header and list of transactions [11, 12]. Block header contains Metadata of the block. Every block in the blockchain inherits from the previous block. The block is constructed using mining statistics. This measure has to be enough complicated to make them tamper-proof and based on the following formula:

$$H_k = \mathrm{Hash}(H_{k-1}|T|\,|nonce|)$$

where T and nonce can be obtained by solving the consensus mechanism. The hash of the current block can be calculated by the Hash of previous block Hash value, transaction root Hash value. Block is also containing the Merkle Tree Root.

2.2 Blockchain Working Process

The Blockchain working process is simply described by four steps [13], as indicated below:

1. The sending node records new data and generates the necessary hash and broadcast that to the network;
2. The receiving node checks the message's hashes and the content if the message is correct, then it will be stored to its block; this process is generally done through what is called a "proof effort", e.g., proof of work (PoW) or proof of stake (PoS) or other model depending on the type of blockchain in use. However, the dominant ones are PoW and PoS [14].

When the majority of the nodes store the block, and it builds on it and moves to the next one. Blockchain working process is shown in Fig. 4.

With the PKI being at the forefront of the blockchain's architecture; the consensus is what makes it all work together. The consensus is what allows the different nodes to agree to the policies in the network. In the case of Bitcoin and many other blockchain technologies like Ethereum (as to the date of writing), the consensus is called the Proof of Work; it is worth noting here that there are many types of consensuses such as Proof of Stake [15].

Proof of Work (PoW)—as highlighted is the software algorithm that maintains both the safety as well as the transparency on the blockchain. In the case of Bitcoin, It uses SHA—256 hash functions to operate. Miner nodes operating on the blockchain consider 10 min worth of bitcoin-based. Blockchain activity and then encode those

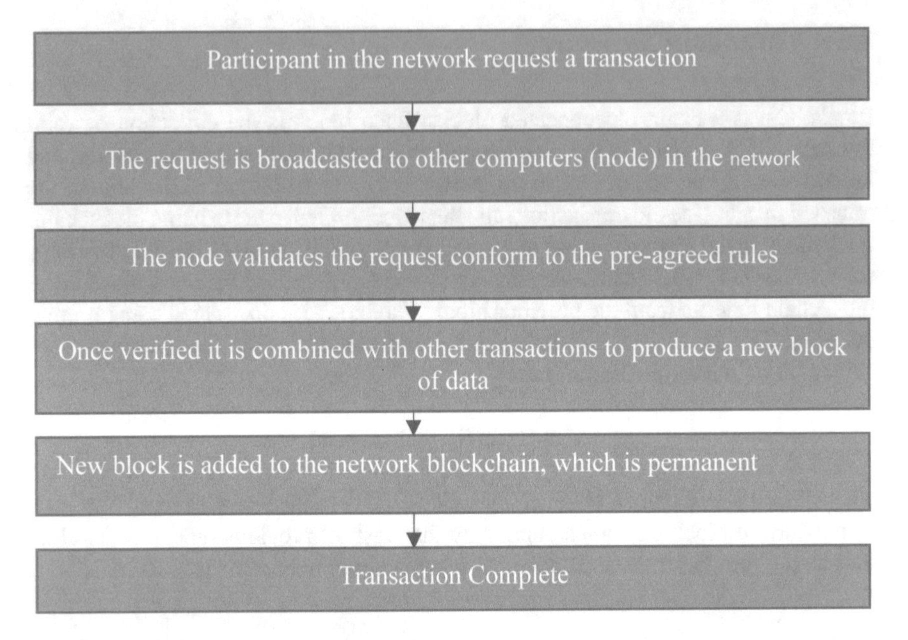

Fig. 4 Blockchain structure

transactions in the form of block. Mathematical solutions are then applied to create a hash value for that block. By doing this, the mining node has not only recorded all the transactions occurring during that time but has also indicated to the more extensive Blockchain network that the work required to hash the transactions has been performed. In return, the Blockchain network provides the mining node with a fixed amount of bitcoins [16].

Making the 'hash' is a crucial process in the proof of work. Here, the researchers [17] said that a proof of work incorporates the same operational principles as a Completely Automated Public Turing test to tell Computers and Humans Apart or CAPTCHA where any potential users have to go through and pass a test in order to access services. The proof of work leverages on the computational power of the Blockchain network, preventing any attempt to interfere or tamper with the blocks.

Proof of Stake (PoS)—This model came to life to tackle the drawbacks of PoW. As we have seen with the PoW the more hashing power you have the likely you will be rewarded; This made for what is called in the crypto community a "whale" where an entity owns most of the power of coins. The entity with the most potent machine will get to control the network. That is why PoS was proposed. In the simplest form PoS is when the users in the Blockchain network buy tokens (the asset-specific to the system like bitcoins in the case of the Bitcoin network) that permit them to conduct transactions involving decision making events and the future of the Blockchain network [18]. Users use the tokens to indicate that they have sufficient balances to participate. The authors [19] said that this is more environmentally friendly than PoW functions as they do not require as much computational power as mining operations.

Hence the fees for transacting using PoS are lower. However, the PoS mechanism has been criticized as being excessively centralized. While [20] pointed out that PoS permits broader user participation amongst users, is less susceptible to centralization.

2.3 *Blockchain Types*

Blockchain technologies divided into three types: (1) Public Blockchain(permissionless); (2) Consortium Blockchains and (3) private Blockchain(permissioned) [13, 14].

2.3.1 Public Blockchain

A Blockchain was designed to securely cut out the middleman in any exchange of asset scenario. It does this by setting up a block of peer-to-peer transactions. Each transaction is verified and synced with every node affiliated with the blockchain before it is written to the system. Until this has occurred, the next transaction cannot move forward. Anyone with a computer and internet connection can set up as a node that can sync with the entire blockchain history. Although this redundancy has it's advantages such as making public blockchains extremely secure; it is also a contributing reason to making then slow and somewhat wasteful [21]. A public blockchain has several benefits, such as:

1. Every transaction is public, and users can maintain individual anonymity;
2. Provides decentralization and becomes an excellent advantage for situations where a network needs to be decentralized;
3. Full transparency of the ledger; and
4. Faster, secure, and less expensive than the accounting systems and methods used today in the finance industry, however, the costs are higher, and speeds are slower than on a private chain. This also means that Public Blockchains allow any person with an internet connection to participate in the verification of transactions process and set themselves as a node [22].

2.3.2 Private Blockchain

Private blockchain lets the middleman back in, to a certain extent. The user writes and verifies each transaction allowing for greater efficiencies, and significant speed on private blockchains [14]. However, here is the main argument.

The company can choose who has read access to their blockchain's transactions, allowing for greater privacy than a public blockchain. A private blockchain is a better fit for more traditional business. This means that the private chains allow one party to have full control, and they will select a few nodes that are predetermined.

2.3.3 Consortium Blockchain

A Consortium blockchain is partially private. Consortium blockchain platforms have many of the same advantages as a private blockchain but operate under the leadership of a group instead of a single entity. This platform would be great for organizational collaboration. A consortium chains provide many of the same benefits of the private blockchain (efficiency and transaction privacy, etc.), without consolidating power with only one party; Hyperledger is an example [13]. In summary, every Blockchain network has different rules regarding what kind of assets it trades, and under which conditions trading takes place. Those rules encoded into its software called the consensus. The node in the Blockchain network is every device running the Blockchain software and connected to the network [1, 23].

As mentioned earlier, blockchain classified into three options which are public blockchain, private blockchain, and consortium blockchain.

2.4 Blockchain Characteristics

The blockchain technology made through different existing technologies such as cryptography, mathematics, algorithms, and distributed consensus algorithms [22, 24, 25]. As such, the Blockchain has six key characteristics [13].

1. *Decentralization*. It means that the Blockchain does not have to rely on a single centralized node which functions as a master node. Each node can record, store, and update the ledger. Together they form the blockchain community of peer-to-peer nodes.
2. *Transparency*. The block's data recorded by each node and distributed among other connected nodes are visible to each node, thus creating openness among connected nodes.
3. *Open Source*. Most Blockchain systems are open to anyone, allowing anybody to modify the code and technology in ways that best suit their needs. However, this does not mean that anyone can edit a running blockchain solution. Making any modification to a working solution means connected nodes agree to accept the change, and it is only valid when connected nodes adopt the change.
4. *Autonomy*. As there are connected nodes, any changes happen once the majority of nodes accept the change. It enforces good deeds from different nodes making changes or intervention useless; other nodes will easily detect any attempt at making any change.
5. *Immutable*. The records will be preserved forever, and cannot be changed unless someone can take control of more than 51% nodes at the same time (i.e., a simple majority).
6. *Anonymity*. Data is hashed and shared; being hashed makes transactions somewhat anonymous.

3 Privacy Requirements for Blockchain

In the context of Blockchain Technology, privacy and confidentiality mean that the data written to the blockchain and the identities of the parties involved are protected [26]. Privacy and confidentiality in the blockchain still a challenge and open issue [27, 28]. This claim also mentioned by several authors such as [19, 29, 30]. Privacy was pointed out as a problem in the original paper of Bitcoin conducted by [22] as there were no ways to protect the privacy of users. With a public blockchain, the information stored in public ledger. And the transaction contains various information such as the ID of the previous transaction, timestamp, participants address, trade values, and signature of its sender [31]. Hence, there is a possibility of tracing the transaction to extract the users' physical identities or other additional information by data mining [16]. Privacy in the blockchain divided into two types which are (1) privacy of information and (2) privacy of the party [19]. Privacy of information is related to the content of the message posted to the blockchain. The party may wish to hide the content of the message from network members. While the privacy of the party is related to the identity of the party, who will be involved in the transaction in the Blockchain [19]. The privacy requirements in the blockchain are studied by several researchers [11, 19, 32, 33]. Hence, to protect privacy, the following requirements should be considered.

The content of the transactions should be only known to their partakers;

- Transaction details are not visible to unauthorized third parties and the world at large unless one of the counterparties has chosen to reveal that information; and
- Transaction details cannot be collected, analyzed, or matched with "off-Blockchain" metadata to reveal any information about counterparties or transaction details. By this, our definition encompasses the use of graph analysis, pattern matching, and machine learning to construct a profile of a counterparty based on the activities associated in the ledger [27, 28, 34]. The blockchain needs to satisfy several requirements to protect privacy [35], and as follows:

  1. The links between transactions should not be visible or discoverable.
  2. The content of transactions is only known to their partakers [16].
  3. The private or permissioned blockchain could set an access control policy. It gives complete transparency of the blockchain data is not a problem.

The privacy requirements should be considered on two factors [16, 36, 37] and as follows:

1. Identity Privacy: which means intractability between the transaction scripts and the real identities of their partakers, as well as the transactional relationships between users. Even if users apply random addresses (or pseudonyms) when acting in the blockchain, they can only provide limited identity privacy.
2. Transaction Privacy: it means that specified users can only access the transaction contents. Transaction privacy is the primary concern in the public blockchain. In the next section, the security and privacy issues in the blockchain explored.

Various measures and techniques that can be used to achieve the security and the privacy at each layer of blockchain investigated as well.

4 Security and Privacy Issues in Blockchain

Since all transactions in public blockchain are visible and open in the network, so the blockchain is mainly vulnerable to leakage of transactional privacy [11]. The critical evaluation parameter in any blockchain is how well the conditions of security and privacy meet the requirements of blockchain. Hence, analyzing the security and privacy issues of blockchain become a valuable research area. Security is defined based on three main components which are confidentiality, integrity, and availability. In blockchain context, privacy means limits the access to the information through a set of rules. Integrity means the information is accurate and trustworthy, and availability means that the information is grantees to be accessed by authorized people. Privacy is defined based on two components data privacy and user privacy [38].

As mentioned earlier, blockchain architecture includes several layers. Thus identifying the challenges that occur in each layer would be very important to be taken into consideration. Encryption measure can be used to achieve confidentiality in three layers which are a smart contract, network, and data layer [38]. Two measures can be used to achieve integrity, which is the Message Authentication Code (MAC) and Signature Scheme. MAC used for achieving the integrity in three layers of blockchain, which are a smart contract, network, and data layer.

Signature Scheme can be used to achieve integrity in both transaction and consensus layers [38]. Availability made by various measures such as consensus, access control, and protocols. Data privacy-preserving computation measure adopted for achieving the data privacy and user privacy at the smart contract layer.

Access control measure used for achieving the data privacy, while blind signature and ring signature used for achieving the anonymity (user privacy) at the consensus layer. Zero-Knowledge Proofs and Mixing measures used for achieving both data privacy and user privacy at the transaction layer. Access control measure used for achieving the data privacy at the data layer, while IP Anonymity measure used for achieving the privacy of user at the network layer [38]. Table 1 summarizes the cryptographic measures that which adopted for achieving security and privacy of information subjected to the blockchain layers.

5 Privacy and Security Measures Used in Blockchain

In this section, we provide a detailed discussion on a selection of techniques that can be leveraged to enhance the security and privacy of existing and future blockchain systems.

Table 1 Cryptographic measures subject to each layer

Blockchain layer	Confidentiality	Integrity	Availability	Data privacy	User privacy
Smart contract	Encryption	MAC	–	Data privacy preserving computation	Identity privacy preserving computation
Consensus	–	Signature scheme	Consensus	Access control	Blind or ring signature
Transaction	–	Signature scheme	Access structure of transactions	Zero-knowledge proofs, mixing techniques	Zero-knowledge proofs
Network	Encryption	MAC	Protocols	–	IP anonymity
Data	Encryption	MAC	Access control	Access control	–

5.1 Mixing

Mixing measures was proposed by Chaum [39]. It aims to hide the identity of users as well as the content of the communication. The architecture of mixing service is clearly stated in Fig. 5.

Here the explanation of mixing by an example. Assume that we have two entities sender and receiver and a message M is prepared to be delivered at address R. The message will be encrypted with the receiver public key KR and appending the address R. Then the intermediary's public key KI is encrypted with the result and based on the following formula:

$$K_I(r_0,\ K_R(r_1,\ M),\ R) \rightarrow K_R\left(r_{1,}M\right),\ R$$

where r_0 and r_1 random numbers which ensure that no message is transferred more than once s we mentioned in the first place, Bitcoin's blockchain doesn't guarantee

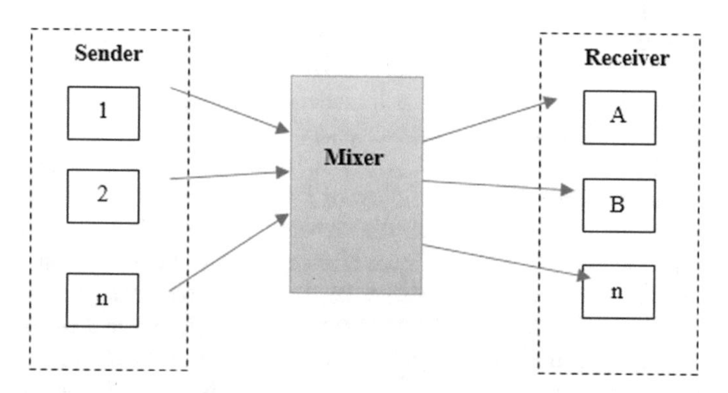

Fig. 5 Mixing service architecture

obscurity for users: transactions use onymous addresses and may be verified publicly, so anyone will relate a user's dealing with her alternative transactions by an easy analysis of addresses she utilized in creating bitcoin exchanges. Additional seriously, once the address of the dealing is coupled to the real-world identity of a user, it's going to cause the outflow of all her transactions. Thus, admixture services (or tumblers) was designed to forestall users' addresses from being coupled. Mixing, literally, it's a random exchange of user's coins with alternative users' coins, as a result, for the observer, their possession of coins is obfuscated. However, these mixing services don't offer protection from coin thieving [40].

Mixcoin was planned by Bonneau et al. in 2014, that provides anonymous payment in Bitcoin and bitcoin-like cryptocurrencies. To defend against passive adversaries, Mixcoin extends the namelessness set to permit all users to combine coins at the same time. To defend against active adversaries, Mixcoin provides namelessness the same as ancient communication mixes. Additionally, Mixcoin uses associate degree responsibleness mechanism to observed stealing, and it shows that users can use Mixcoin rationally while not stealing bitcoins by orienting incentives [41].

5.2 Anonymous Signatures

This section will dive in discussing the two most important and typical anonymous signature schemes which are group signature and ring signature.

Group signature is a cryptography theme which firstly proposed by [42]. Given a bunch, any of its members will sign a message for the whole cluster anonymously by exploitation her personal secret key, and any member with the cluster's public key will check and validate the generated signature and ensure that the signature of some group member is employed to sign the message. The method of signature verification reveals nothing regarding verity identity of the signer except the members of the cluster. Cluster signature encompasses a group manager who manages adding group members, handling the event of disputes, together with revealing the first signer. Within the blockchain system, we have a tendency to conjointly would like a licensed entity to form and revoke the cluster and dynamically add new members to the group and delete/revoke the membership of some participants from the group. Since the group signature needs a bunch manager to line up the group, the cluster signature is appropriate for syndicate blockchain.

Ring signature was proposed by [43] which shown in Fig. 6. It can succeed anonymous through linguistic communication by any member of cluster users. The term "ring signature" originates from the signature algorithmic program that uses the ring-like structure. The ring signature is anonymous if it's troublesome to work out that member of the cluster uses his/her key to sign the message. Ring signatures take issue from cluster signatures in 2 principal ways: 1st, during a ring signature theme, the $64,000 identity of the signer can not be discovered within the event of a dispute, since there's no cluster manager during a ring signature. Second, any users will group A "ring" by themselves while not further setup. Thus, the ring signature

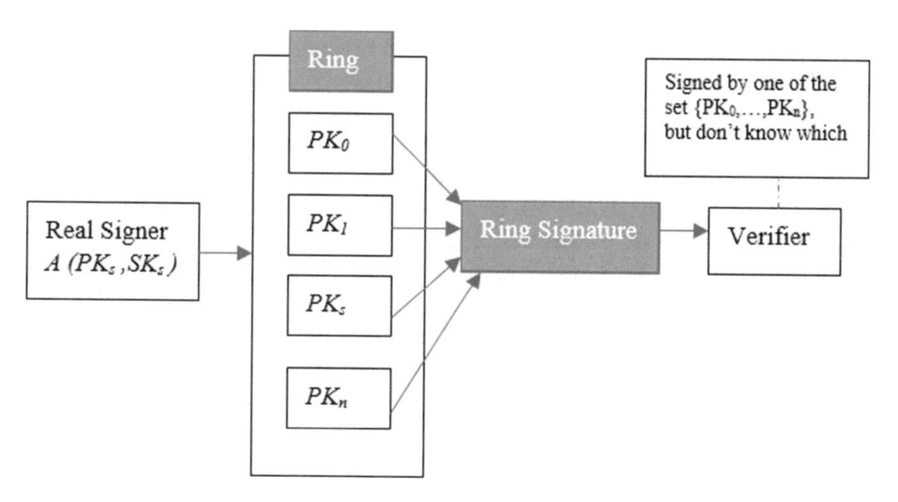

Fig. 6 Ring signature process

is applicable to the general public blockchain. One in every one of the standard applications of a hoop signature is CryptoNote [44]. It adopts ring signature to cover the association between the sender's addresses of transactions. Additional exactly, CryptoNote constructs the sender's public key with many alternative keys, so it is not possible to spot World Health Organization truly sent (signed) the dealing. Thanks to the utilization of ring signature, if the quantity of ring members is n, then the likelihood that AN opponent might with success guess a true sender of dealing is 1/n.

Here is the working process of ring signature measure [43]. User A selects a set of users ($User_0$, $User_1$, $User_n$) and creates a ring. Each user has a pubic key from standard signature scheme such as ECDSA. User A signs a message with his/her private key (SKs) and all the public keys (PK_0, PK_1, PK_n) of the members in the ring. The verifier knows that the message signed by one from the set but does know the real singer. Hence, the ring signature provides anonymity for the signer.

5.3 Homomorphic Encryption (HE)

Homomorphic cryptography (HE) is a powerful cryptography. It will perform sure kinds of computations directly on ciphertext and make sure that the operations performed on the encrypted knowledge, once decrypting the computed results, can generate identical results to those performed by constant operations on the plaintext. There are many parts of homomorphic cryptosystems [45, 46]. One will use homomorphic cryptography techniques to store knowledge over the blockchain with no vital changes within the blockchain properties. This ensures that the information on the blockchain is encrypted, addressing the privacy issues related to public blockchains. The employment of homomorphic cryptography technique offers privacy protection and permits prepared access to encrypted knowledge over public blockchain for auditing and different functions, like managing worker expenses. Ethereum sensible contracts give homomorphic cryptography on knowledge hold on in blockchain for larger management and privacy.

The working process of HE will be explained by the following scenario.

A and *B* are two main parameters. *A* has secret values (x_1, x_2,..., x_n) and *B* has a function *F*(.) In order for A and B to calculate $F(x_1,...,x_n)$ together without leaking the secret values.

E(.)/*D*(.) are the set of the homomorphic encryption system. *A* can send encrypted inputs E(x_1),...,$E(x_n)$*to B*. After that, a normal computation on the encrypted text should be performed and the result send to *A*. *A* will get$f(x_1,...,x_n)$ after the decryption. The privacy of blockchain will be protected in the implementation of homomorphic cryptographic including both the Pedersen commitment scheme and Paillier cryptosystem [16].

5.4 Attribute-Based Encryption (ABE)

Attribute-based encryption (ABE) is a cryptographic method. The conception of attribute-based cryptography was planned in 2005 [47] with one authority. Since then, variety of extensions are planned to the baseline ABE, as well as ABE with multiple authorities to come up with users' personal keys together [48, 49]. ABE schemes that support impulsive predicates [50, 50]. Attribute-based cryptography is incredibly powerful nonetheless few applications thus far deploy it thanks to the shortage of understanding of each core ideas and economical implementation. ABE has not nonetheless been deployed in any type on a blockchain for the data processing thus far. In 2011, a localized ABE theme was planned [49] to use ABE on a blockchain. For instance, on a blockchain, permissions may well be delineated by the possession of access tokens. All nodes within the network, that have an exact token issued to them, are going to be granted access to the special rights and privileges related to the token. The token provides a method of following who has bound attributes Associate in Nursingd such tracking ought to be worn out an recursive and consistent fashion by the approved entity that distributes the token. Tokens are often viewed as badges that represent attributes or qualifications and may be used as non-transferable quantifiers of name or attributes.

5.5 Secure Multi-party Computation

The multi-party computation (MPC) model defines a multi-party protocol to permit them to hold out some computation together over their non-public knowledge inputs while not violating their input privacy, such that associate degree soul learns nothing concerning the input of an authentic party however the output of the joint computation. The primary large-scale preparation of MPC was in 2008 for associate degree actual auction downside in Denmark [51]. In recent years, MPC has been utilized in blockchain systems to shield users' privacy [52]. Designed and enforced secure multiparty computation protocols on the Bitcoin system in 2014. They made protocols for secure multiparty lotteries with none sure authority. Their protocols are ready to guarantee fairness for honest users despite however dishonest ones behave. If a user violates or interferes with the protocol then she becomes a loser and her bitcoins are transported to the honest users. A decentralized SMP computation platform, referred to as Enigma, is projected in 2015 by [53]. By exploitation a sophisticated version of SMP computation, Enigma employs a verifiable secret sharing theme to ensure the privacy of its process model. Also, Enigma encodes shared secret knowledge employing a changed distributed hash table for economical storage. Moreover, it leverages associate degree external blockchain as a corruption-resistant recording of events and also the regulator of the peer-to-peer network for identity management and access management. Like the Bitcoin system, Enigma provides autonomous management and protection of private knowledge whereas eliminating the requirement and dependency of a sure third party.

As mentioned earlier, MPC is a cryptographic protocol that used for emulating a trusted party. MPC would be very benefit to be used in system with no trusted parties. MPC has two main goals which are the correctness and security [54]. Its working can be shown based on the following example.

Let say P_1, P_2,...,P_n are mutually suspicious. Each party has a secret input (x_1, x_2, x_n). Hence, the joint function y can be computed as follows:

$$y = f(x_1, x_2, \ldots, x_n)$$

The correctness goal will be achieved through that everyone can computes $y = f(x_1, x_2,\ldots,x_n)$ and the security goal will be achieved through that nothing but the output is revealed.

5.6 Non-interactive Zero-Knowledge (NIZK) Proof

Zero-knowledge proofs are another cryptographic technology that has powerful privacy-preserving which was proposed in the early 1980s [55, 56]. The concept of this protocol was proposed by Blum. The goal of NIZK is to preserve the privacy. It aims to provide the prove the correctness of data without leaking addition information. A zero-knowledge protocol can be explained by an example.

Let say we have two parties the prover P and the verifier V. The prover has a statement which he wants to prove to verifier. Hence, the working process of zero-knowledge proof will be based on the following steps:

1. The prover will compute a proof for the statement and then send the proof to the verifier.
2. The verifier will choose a question and send it to the prover.
3. The prover then calculates the answer for the question and send it to the verifier. Hence, the verifier by using the answer will check whether Prover really knows the statement. Figure shows an interactive zero-knowledge protocol between prover and verifier.

NIZK proof system can be represented according the following formulas [16].

(P, V) are the prover and verifier, the NIZK proof system for language L when $L \subseteq NP$ with k where k is a security parameter and if it meets two properties which are completeness and soundness.

The completeness occurs for any input $x \in L$ and its witness w and polynomial $p(.)$

$$P_r[V(R, x, P(R, x, w)) = 1] \geq 1 - 1/P(x)$$

The soundness occurs for any input x $\notin$ L and algorithms P^* and polynomial $p(.)$

$$P_r\left[V\left(R, x, P^*(R, x,)\right) = 1\right] < 1 - 1/P(x)$$

5.7 The Trusted Execution Environment (TEE) Based Smart Contracts

TEE is an execution atmosphere if it provides a totally isolated environment for application execution, that effectively prevents different software package applications and operative system(s) from meddling with and learning the state of the applying running in it. The Intel software package Guard eXtensions (SGX) maybe a representative technology to implement a TEE. As an example, Ekiden [18] may be an SGX primarily based resolution for confidentiality-preserving sensible contracts. Ekiden separates computation from the accord. It performs sensible contract computation in TEEs on calculating nodes off-chain, then uses a foreign attestation protocol to validate the execution correctness of calculating nodes on-chain.

Intel® SGX measure is a hardware-based solution which provides data protection. It is a platform with built-in CPU instructions that permit the access to the data. Access to the data will be denied or disabled if the code is altered or tampered.

5.8 Discussion

Three main points have to be taken into consideration in order to achieve security and privacy in the blockchain.

(1) No single technology may be a cure for the security and privacy of Blockchain. (2) There's no technology that has no defects or is ideal altogether aspects. Once we add new technology to a posh system, it perpetually causes alternative issues or new form(s) of attacks and (3) there's perpetually a trade-off between security-privacy and potency. We must always advocate those techniques that improve the protection and privacy of blockchain It is clearly stated in the previous sections that blockchain technology classified into three types: (1) public blockchain, Private blockchain and Consortium blockchain. Privacy and confidentiality are still a challenge in the blockchain. Hence, this research comes into address the privacy issues and analyzed the solutions that would be used in order to preserve and maintain the privacy. This research addresses several measures that can be used for preserving the privacy in blockchain and as shown in Table 2.

It is very important to maintain the privacy at all levels such as data level, transaction level and network level. Privacy can be maintained very well in permissioned blockchain because of that (1) running private blockchain is easier that public blockchain, (2) easy to change the rules and revert the transactions, (3) the validators are known so any risk of a 51% attack arising from some miner collusion does not apply, (4) transactions are cheaper and since they only need to be verified by a few nodes that can be trusted to have very high processing power, and do not need to be verified by ten thousand laptops, (5) Nodes can be trusted to be very well-connected and (6) private blockchain can provide a greater level of, well, privacy because the read permissions are restricted. Privacy in private blockchain can be maintained using different measures and it is based on the platform. Hyperledger fabric is a private permissioned blockchain. From all the above privacy techniques which mentioned earlier, Hyperledger uses attribute-based encryption of data in which can restrict data to a user based on user's attribute. Hyperledger fabric can also zero knowledge proof in which verifier can verify issuer without getting access to issuer data. As mentioned above that in Hyperledger fabric we can restrict access between user of same organizations we can create private data side DB in which only those user will have access which are linked to the transaction. Hyperledger fabric ledger consist of world state (Database) and Transaction log (Blockchain), there is public ledger for all permissioned participants but in public ledger there are only hashes not actual data. World state is maintained so reading data doesn't involve traversing the entire blockchain. Each peer can recreate the world state from the transaction log. Private data and Attribute Based Encryption together give enough flexibility to model a non-trivial business process without revealing confidential information. Hyperledger fabric adds certain layers of privacy for user's data, First, it gives access to only permissioned user of the network, then can secure user's actions using attribute based encryption. Moreover, it uses zero knowledge proof where authentication is required between user without giving data of one user to another. In order to maintain the privacy at

Table 2 Privacy measures in blockchain

Techniques	Application	Advantages	Disadvantages
Mixing	MixCoin	It is very helpful in preventing the users' addresses from being linked	There might risk of leakage of user privacy due to the centralized services
Group signature	JUCIX	The ability to hide the signer identity among a group of users	Trusted third party is needed to act as a manager
Ring signature	CryptoNote, Ethereum	The ability to hide the signer identity among a group of users and hence trusted the third party is not required	The signer identity can not be revealed in the event of a dispute
ABE	None	Data confidentiality and fine-grained access control can be achieved simultaneously	Need to resolve the issuance and revocation of attribute certificate in the distributed environment
HE	Ethereum	Privacy-preserving can be achieved by performing a computation on the ciphertext	The computational efficiency of the complex function is very low
SMPC	Engima	The ability to carry out some computation through multi-party without violating the input privacy	Efficiency is low in the complex functions
NIZK	Zcash	Users can prove their balance easily without reveling the account balance	Less efficient
TEE based solutions	Ekiden, Enigma	The privacy of smart contract can be protected by running them in TEE	Need to resolve the attacks on SGX

data level, end to end encryption can be used while preserving the privacy at network level could be through network level configurations as well as using private channels. Preserving the privacy at truncation level could be through implementing access control measures. In case of public blockchain, and based on the analyzing we made for the existing measures and techniques, different encryption schemes for both identity privacy and transaction privacy can be used and based on the use case. Ethereum adopts a couple of measures for maintaining the privacy which are Zero Knowledge-based and Mixers.

From what has been presented, it can conclude that the measures of privacy can be successfully employed based on the blockchain platform and the use case.

6 Performance Evaluation of Blockchains

Ethereum and Hyperledger are the most dominant blockchain platforms. Smart contract and Crypto-currency are an example of the application in Ethereum. The smart contract is executed in Ethereum Virtual Machine (EVM) using Solidity Language and Ethash (PoW) consensus algorithm. In Hyperledger, the smart contract is executed using Dockers and Golang and Java languages, and Practical Byzantine Fault Tolerance (PBFT) consensus algorithm [14]. Authors [57] conducted a performance analysis of Ethereum and Hyperledger.

Hyperledger Fabric consistently performs better than Ethereum both in term of throughput and latency. This research will dive in more details regarding the measures of the private security mechanism of Hyperledger Fabric. The privacy protection measures of Hyperledger Fabric will be divided into four measures:

1. Using symmetric cryptography and zero-knowledge proof. This is for several reasons such as separating the transaction data from on-chain records and protecting privacy from the underlying algorithm.
2. Using the digital certificate management service. This is for guarantees the legitimacy of the organization on the blockchain.
3. Using the design of multi-channel. This is for separating the information between different channels.
4. Privacy data collection. This is for satisfying the need for the isolation of privacy data between different organizations within the same channel.

The channel and privacy data collection are the most typical methods. The channel is dedicated to allowing the data on the channel to be isolated separately and to the blockchain privacy protection. The ledger is shared by the peer on the same channel and the recognition of the channel is needed to be obtained by the transaction peer before it can join the channel and transact with others. The private data collection (PDC) is a group of organizations that are permitted to store private data on a channel. The data stored contains the private data and the hash value of the private data [58].

In the Hyperledger Fabric, the processing of privacy data is divided into two scenarios: new channels are required once the whole dealings and ledger should be unbroken strictly confidential to the skin members of the channel; when the transaction info and ledger have to be compelled to be shared among some organizations, a number of them are going to be able to see all the dealings information, alternative organizations have to be compelled to recognize the prevalence of this dealings to verify the genuineness of the transaction, a non-public information assortment ought to be established during this case. Additionally, as a result of non-public information is propagated through peer-to-peer instead of block, the privacy information assortment is employed once the dealings data should be confidential for the sorting service peer. The blockchain dealings method involving the privacy information assortment is as follows [59, 60].

Blockchain transaction process involving the privacy data collection based on several steps and as follows:

1. The offer request is submitted by the client application to call the chain code function to the endorsement peer of the private data set authorization and through the provisional domain, the private data is sent.
2. The transaction is simulated by the endorsement peer and the private data is stored in a local temporary repository in the peer. The gossip protocol is used by the endorsement peer in order to disseminate the private data to the authorized peer.
3. The public data is returned by the endorsement peer including the hash value of the private data key-value pair.
4. The transaction is submitted to the sorting service peer by the client application and then distributing the sorting result to each block.
5. The authorized peer can use the collection policy when submitting a block in order to determine if it is authorized to view private data.

7 Conclusion

In this research, we identified and discussed the cryptographic algorithms used in the blockchain. In addition to that, we identified the privacy requirements for the blockchain as well as the privacy techniques that can be used to manage the privacy in the blockchain. Our findings show that there are many techniques which can be used for enhancing the privacy in the blockchains. Finally, we discussed the privacy process in the Hyperledger platform and the measures that can be used for ensuring its privacy.

Acknowledgements Authors would like to sincerely thank Universiti Utara Malaysia (UUM), International Islamic University Malaysia (IIUM), Malaysia and Ministry of Higher Education, Iraq for supporting this research.

References

1. Grech, A. and Camilleri, A. F.: Blockchain in Education. In: Inamorato dos Santos, A. (ed.) EUR 28778 EN (2017). https://www.doi.org/10.2760/60649
2. Ackerman, A.,Chang, A., Diakun-Thibault, N., Forni, L., Landa, F., Mayo, J., van Riezen, R.: Blockchain and Health IT: Algorithms, Privacy and Data (August 8, 2016). Project PharmOrchard of MIT's Experimental Learning "MIT FinTech: Future Commerce.", White Paper August 2016. Available at SSRN: https://ssrn.com/abstract=3209023

3. Duan, Z, Mao, H., Chen, Z., Bai, X., Hu, K., Talpin, J.-P.: Formal modeling and verification of blockchain system, vol. 86, pp. 231–235 (2018)
4. Wu, J., Tran, N.K.: Application of blockchain technology in sustainable energy systems: an overview. Sustain **10**(9), 1–22 (2018)
5. Cui, G., Shi, K., Qin, Y., Liu, L., Qi, B., Li, B.: Application of block chain in multi-level demand response reliable mechanism. In: 2017 3rd International Conference on Information Management (ICIM), pp. 337–341 (2017)
6. Fukumitsu, M., Hasegawa, S., Iwazaki, J., Sakai, M., Takahashi, D.: A proposal of a secure P2P-type storage scheme by using the secret sharing and the blockchain. In: Proceedings of the International Conference on Advanced Information Networking and Applications (AINA), pp. 803–810 (2017)
7. Yuan, Y., Wang, F.Y.: Towards blockchain-based intelligent transportation systems. In: IEEE International Conference on Intelligent Transportation Systems (ITSC), pp. 2663–2668 (2016)
8. Zheng, Z., Xie, S., Dai, H.N., Wang, H.: Blockchain challenges and opportunities: a survey. Work Pap.–2016, December 2016
9. Baliga, A.: Understanding blockchain consensus models. Whitepaper, April, pp. 1–14 (2017)
10. Zheng, Z., Xie, S., Dai, H., Chen, X., Wang, H.: An overview of blockchain technology: architecture, consensus, and future trends. In: Proceeding of 2017 IEEE 6th International Congress on Big Data (BigData Congress), pp. 557–564 (2017)
11. Prashanth Joshi, A., Han, M., Wang, Y.: A survey on security and privacy issues of blockchain technology. Math. Found. Comput **1**(2), 121–147 (2018)
12. Le, T., Mutka, M.W.: Capchain: a privacy preserving access control framework based on blockchain for pervasive environments. In: Proceedings of 2018 IEEE International Conference on Smart Computing (SMARTCOMP), pp. 57–64 (2018)
13. Lin, I.-C., Liao, T.-C.: A survey of blockchain security issues and challenges. Int. J. Netw. Secur. **1919**(55), 653–659 (2017)
14. Dinh, T.T.A., Wang, J., Chen, G., Liu, R., Ooi, B.C., Tan, K.-L.: Blockbench: a framework for analyzing private blockchains. In: Proceedings of the 2017 ACM International Conference on Management of Data. ACM (2017)
15. Fabian, B., Ermakova, T., Krah, J., Lando, E., Ahrary, N.: Adoption of security and privacy measures in bitcoin–stated and actual behavior (2018). Available at SSRN:https://ssrn.com/abstract=3184130
16. Feng, Q., He, D., Zeadally, S., Khan, M.K., Kumar, N.: A survey on privacy protection in blockchain system. J. Netw. Comput. Appl. **126**, 45–58 (2019)
17. Duan, B., Zhong, Y., Liu, D.: Education application of blockchain technology: learning outcome and meta-diploma. In: Proceedings of the International Conference on Parallel and Distributed Systems (ICPADS), December 2017, pp. 814–817 (2018)
18. Cheng, R., Zhang, F., Kos, J., He, W., Hynes, N., Johnson, N., ... & Song, D.: Ekiden: A platform for confidentiality-preserving, trustworthy, and performant smart contracts. In 2019 IEEE European Symposium on Security and Privacy (EuroS&P), pp. 185–200. IEEE (2019, June)
19. Axon, L., Goldsmith, M., Creese, S.: Privacy requirements in cybersecurity applications of blockchain, vol. 111, 1st edn. Elsevier (2018)
20. Ruffing, T., Moreno-sanchez, P., Kate, A.: CoinShuffle: practical decentralized coin mixing for bitcoin—bookmetrix analysis. In: European Symposium on Research in Computer Security (ESORICS), vol. 8713, pp. 1–15 (2014)
21. Chen, J., Yao, S., Yuan, Q., He, K., Ji, S., Du, R.: CertChain: public and efficient certificate audit based on blockchain for TLS connections. In: Proceedings of the IEEE INFOCOM, April 2018, pp. 2060–2068 (2018)
22. Nakamoto, S.: Bitcoin: a peer-to-peer electronic cash system, p. 9. Www.Bitcoin.Org (2008)
23. Turkanovic, M., Holbl, M., Kosic, K., Hericko, M., Kamisalic, A.: EduCTX: a blockchain-based higher education credit platform. IEEE Access **6**, 1–20 (2018)
24. Gervais, A., Karame, G.O., Wüst, K., Ritzdorf, H.: On the security and performance of proof of work blockchains Vasileios Glykantzis Srdjaň Capkun. Bitcoin.org (2017)

25. Garay, J.A.: The bitcoin backbone protocol : analysis and applications the bitcoin backbone protocol : analysis and applications, June 2017, pp. 1–44 (2015)
26. Yang, D., Gavigan, J., Hearn, Z.W.: Survey of confidentiality and privacy preserving technologies for blockchains, pp. 1–32 (2016)
27. Stuart, P.: Confidentiality in Private Blockchain (August 8, 2016). Project "Kadena: Kuro - Private Blockchain.", White Paper August 2016. Available at SSRN:https://www.kadena.io/
28. Chang, P., Yang, C., Yang, C., Hwang, M.: An academic transcript system embedded with blockchains (2018)
29. Ouaddah, A., Elkalam, A.A., Ouahman, A.A.: Europe and MENA Cooperation Advances in Information and Communication Technologies, vol. 520, pp. 523–533. Springer, Cham (2017)
30. Ikeda, K.: Security and privacy of blockchain and quantum computation, 1st ed., vol. 111. Elsevier (2018)
31. Bhowmik, D., Feng, T.: The multimedia blockchain: a distributed and tamper-proof media transaction framework. In: International Conference on Digital Signal Processing (DSP), 2017 August, November 2017
32. Fan, K., Ren, Y., Wang, Y., Li, H., Yang, Y.: Blockchain-based efficient privacy preserving and data sharing scheme of content-centric network in 5G. IET Commun. **12**(5), 527–532 (2018)
33. Colloquium, J.N., Zrt, B.E.: Blockchain: solving the privacy and research availability tradeoff for EHR data. In: IEEE 30th Jubilee Neumann Colloquium, pp. 135–140 (2017)
34. Ali, A., Afzal, M.M.: Confidentiality in blockchain. Int. J. Eng. Sci. Invent. **7**(1), 50–52 (2018)
35. Wang, R., He, J., Liu, C., Li, Q., Tsai, W.T., Deng, E.: A privacy-aware PKI system based on permissioned blockchains. In: Proceedings of IEEE International Conference on Software Engineering and Service Science (ICSESS) November 2018, pp. 928–931 (2019)
36. Chen, Y., Xie, H., Lv, K., Wei, S., Hu, C.: DEPLEST: a blockchain-based privacy-preserving distributed database toward user behaviors in social networks. Inf. Sci. (NY) **501**, 100–117 (2019)
37. Casino, F., Dasaklis, T.K., Patsakis, C.: A systematic literature review of blockchain-based applications: current status, classification and open issues. Telematics Inform **36**, 55–81 (2018)
38. Raikwar, M., Gligoroski, D., Kralevska, K.: SoK of used cryptography in blockchain (2019)
39. Chaum, D.: Untraceable electronic mail, return addresses and digital pseudonyms. In Secure Electronic Voting, pp. 211–219. Springer, Boston, MA (2003)
40. Zhang, R., Xue, R., Liu, L.: Security and privacy on blockchain, **1**(1) (2019)
41. Bonneau, J., Narayanan, A., Miller, A., Clark, J., Kroll, J. A., & Felten, E. W. (2014, March). Mixcoin: Anonymity for Bitcoin with accountable mixes. In International Conference on Financial Cryptography and Data Security, pp. 486–504. Springer, Berlin, Heidelberg
42. Chaum, D., Van Heyst, E.: Group signatures. In: Lecture Notes in Computer Science (Including Subseries Lecture Notes in Artificial Intelligence and Lecture Notes in Bioinformatics), vol. 547, No. iii, pp. 257–265. LNCS (1991)
43. Wood, G., et al.: How to leak a secret. J. Br. Blockchain Assoc., vol. 2018, November 25, 2016, p. Github site to create pdf, 2016
44. Van Saberhagen, N.: CryptoNote v 2.0. Self-published, pp. 1–20 (2013)
45. Logarithms, D.: A public key cryptosystem and a signature based on discrete logarithms, vol. I, pp. 10–18 (1976)
46. Abidin, A.S.Z., Yusuff, R.M., Bakar, N.A., Awi, M.A., Zulkifli, N., Muslimen, R.: Public-key cryptosystems based on composite degree residuosity classes. In: Lecture Notes in Electrical Engineering (LNEE), vol. 130, pp. 285–299 (2013)
47. Sahai, A., Waters, B.: Fuzzy identity-based encryption BT. In: Advances in Cryptology (EUROCRYPT 2005), vol. 3494, Chapter 27, p. 557 (2005)
48. Chase, M.: Multi-authority attribute based encryption. In: Proceedings of the 4th Conference Theory Cryptography, vol. 4392, pp. 515–534 (2007)
49. Lewko, A., Waters, B.: Decentralizing attribute-based encryption, vol. 2, No. subaward 641, pp. 568–588 (2011)

50. Garg, S., Gentry, C., Halevi, S., Sahai, A., Waters, B.: Attribute-based encryption for circuits from multilinear maps. In: Lecture Notes in Computer Science (Including Subseries Lecture Notes in Artificial Intelligence and Lecture Notes in Bioinformatics), vol. 8043, PART 2, pp. 479–499. LNCS (2013)
51. Bogetoft, P., et al: Secure multiparty computation goes live. In: Lecture Notes in Computer Science (Including Subseries Lecture Notes in Artificial Intelligence and Lecture Notes in Bioinformatics), vol. 5628, pp. 325–343. LNCS (2009)
52. Andrychowicz, M., Dziembowski, S., Malinowski, D., Mazurek, Ł.: Secure multiparty computations on bitcoin. In: Proceedings of the IEEE Symposium on Security and Privacy, pp. 443–458 (2014)
53. Srichaiyo, T., Hjertén, S.: Enigma: decentralized computation platform with guaranteed privacy. J. Liq. Chromatogr **12**(5), 809–825 (2015)
54. Benhamouda, F., Halevi, S., Halevi, T.: Supporting private data on Hyperledger fabric with secure multiparty computation. IBM J. Res. Dev. **63**(2), 1–8 (2019)
55. Goldwasser, S., Micali, S., Rackoff, C.: The knowledge complexity of interactive proof systems. SIAM J. Comput. **18**(1), 186–208 (2005)
56. Essaf, F.: Privacy protection issues in blockchain technology, pp. 124–131 (2019)
57. Pongnumkul, S., Siripanpornchana, C., Thajchayapong, S.: Performance analysis of private blockchain platforms in varying workloads. In: 2017 26th International Conference on Computer Communication and Networks (ICCCN), pp. 1–6. IEEE (July, 2017)
58. Ma, C., Kong, X., Lan, Q., Zhou, Z.: The privacy protection mechanism of Hyperledger fabric and its application in supply chain finance. Cybersecurity **2**(1), 15 (2019)
59. Androulaki, E., et al.: Hyperledger fabric: a distributed operating system for permissioned blockchains. In: Proceedings of the Thirteenth EuroSys Conference, No. 1. ACM (2018)
60. Vukolić, M.: Rethinking permissioned blockchains. In: Proceedings of the ACM Workshop on Blockchain, Cryptocurrencies and Contracts (BCC), pp. 3–7 (2017)

Empirical Evaluation of Blockchain Smart Contracts

Imane Mokdad and Nabil M. Hewahi

Abstract One of the building blocks of our legal and economic systems in society is the indispensable reliance on contracts and trust systems to protect individual rights. Recently smart contracts are becoming prominent parts of various blockchain platforms. The goal of smart contracts is to eliminate the third party and centralized trust systems. Due to recent emergence of smart contracts, there is no well-defined framework that researchers can use to evaluate smart contracts under various blockchain platforms and differentiate between them. In this work, a survey on the prominent smart contract landscape specially those based on blockchain have been conducted. Based on the survey, an evaluation framework to assess smart contracts has been proposed. The framework is a set of criteria based on two major aspects; infrastructure related and development related criteria. The evaluation framework was peer-reviewed for reliability and validity. To measure the applicability of the proposed framework, it has been used to empirically evaluate some of the most prominent smart contract platforms. The results of the empirical evaluation have shown that the Ethereum blockchain smart contract exceeds the others in terms of development tools, resources, and community support. EOS blockchain smart contracts have the best execution speeds, and transaction costs. Lastly, Stellar blockchain has predictability and the best transaction builder to use in smart contract development concerning user friendliness. Recommendations for smart contract developers are provided in light of the research.

Keywords Blockchain · Smart contracts · Evaluation framework · Empirical evaluation

---

I. Mokdad (✉) · N. M. Hewahi
College of Information Technology, University of Bahrain, Zallaq, Bahrain

N. M. Hewahi
e-mail: nhewahi@uob.edu.bh

M. A. Khan et al. (eds.), *Decentralised Internet of Things*, Studies in Big Data 71,
https://doi.org/10.1007/978-3-030-38677-1_3

1 Introduction

To an observer, it may seem that the binding between smart contracts and the blockchain is compulsory; however, this is not the case. The concept of smart contracts existed way before the idea of blockchain was even conceived. The term 'smart contract' was first proposed by Szabo in 1994 [1]. In his paper, Szabo defined smart contracts as "a computerized transaction protocol that executes the terms of a contract". The idea behind it was to convert the clauses of a traditional contract into code that is then embed into hardware or software which enforces the terms of that contract, thereby eliminating the need for third-party entities and trust agencies. In that form, smart contracts were merely a potential prospect and were difficult to implement. In 2008, Nakamoto introduced the Blockchain concept [2]. Due to its ability to provide trust within an untrustworthy distributed environment, it enabled various technologies to come to life. With the availability of that foundational technology, smart contracts became possible. Shortly after, smart contracts were popularized by the Ethereum platform that was released in 2015 [3]. The smart contract concept has taken off since then with proof of concept studies, production implementations, high level distributed applications (DApps), and they became what we know them today. Figure 1 demonstrates a summarized timeline of the history of smart contracts.

As Clack et al. [4] define it, a smart contract is: "An automatable and enforceable agreement. Automatable by computer, although some parts may require human input and control. Enforceable either by legal enforcement of rights and obligations or via tamper-proof execution of computer code." Smart contracts can be developed on top of various blockchain platforms, such as; Ethereum, Hyperledger Fabric, and Corda. The following is an example of a traditional contract: Bob sells Alice a software product license for the cost of 100$. She decides to pay Bob via conventional bank transfer. She begins by filling papers for international money transfer from her account to Bob's account. If the receiving bank does not own an account on the sending bank then intermediary banks will get involved otherwise the transfer would happen directly between the two banks. The process usually takes a few days to finish as shown in Fig. 2. In short, the process involves multiple parties and is highly centralized.

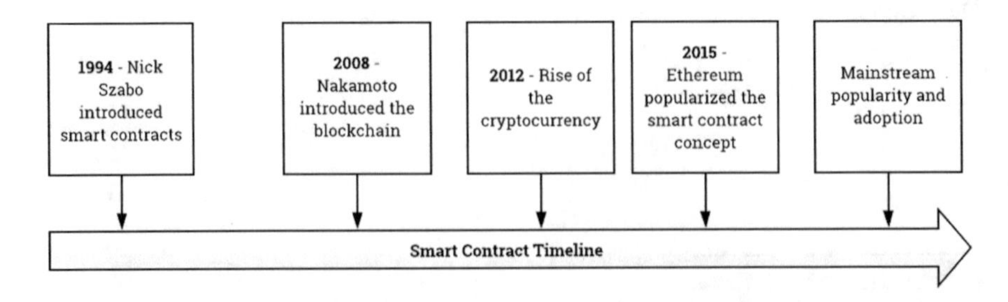

Fig. 1 History of the smart contracts

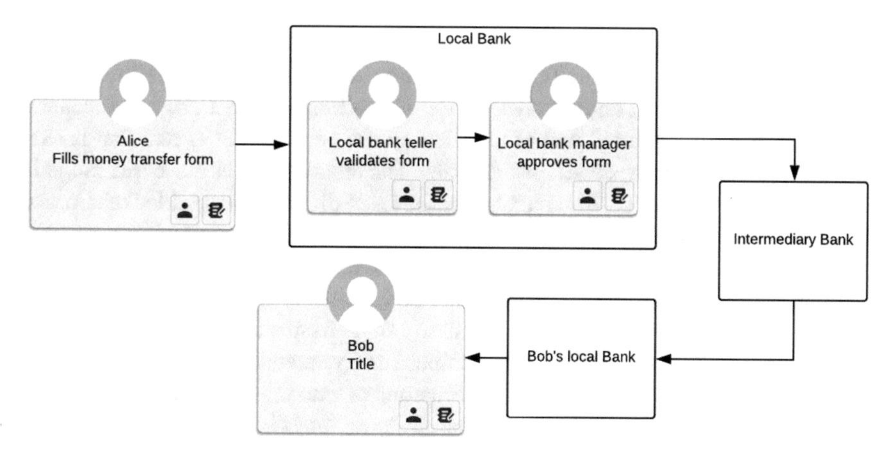

Fig. 2 Alice sends the money over international fund transfer to Bob's account

This scenario is cleverly enhanced eliminating the need for third parties, security attacks, and human error using the blockchain as follows: A smart contract is designed to deduct 0.54 ETH (a blockchain currency with 0.53 ETH of value equivalent to 100$, and hypothetical 0.01 ETH transaction fees) from Alice's account and hold it, then deliver the license key to Alice. Once Alice confirms the license file is valid, the amount is deposited to Bob's account; Bob can reuse the same contract with other customers as well. The whole process is decentralized, provides trust among non-trusting parties, reduces the middleman costs, and finishes in a matter of minutes instead of days given the right gas price (fees) as shown in Fig. 3.

This blockchain smart contract implementation makes transactions transparent, easier, cheaper, and faster [5]. The information is permanently stored on the blockchain, and is free from tampering with evidence stored all around the world.

According to Wang et al. [6], blockchains are constructed of six different layers. The *data layer* that acts as the physical layer of the blockchain. It contains the data, time stamps, blocks, etc. that make up a blockchain. The *network layer* that encapsulates the network topology and network protocols such as P2P. Next is the *consensus layer* that contains the consensus algorithms that pertain to blockchains. The *incentive layer* contains all mechanisms and algorithms that provide network incentives

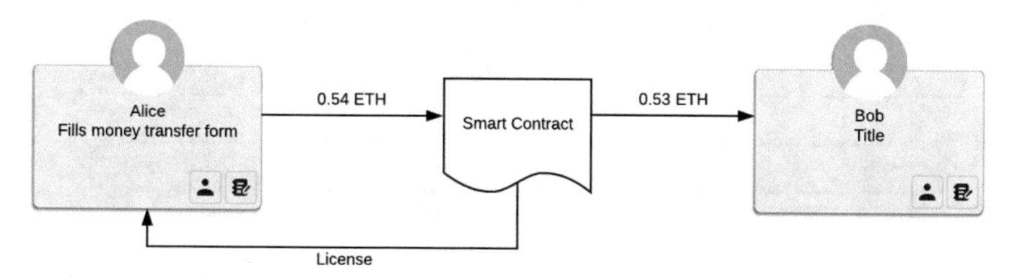

Fig. 3 Alice utilizes the smart contract to purchase her licenses from Bob

and revenue. The *contract layer* encapsulates script code and algorithms that get executed on the blockchain. Finally, the application layer that contains DApps. Smart contracts reside in the contract layer of the blockchain, in the form of procedural program code that is stored and propagated on the network nodes and that is executed when the program conditions are met. The smart contract use-cases range in a wide spectrum from basic contracts that resides directly on the blockchain and only performs financial transfers, to more complex that runs as part of a DApp such as loan systems [7], online voting apps [8], and sharing apps in the likes of Uber and Airbnb [9], to most complex that span an entire society. Figure 4 shows some examples of the use cases. A simple use case can be from a family member creating a transaction to send some money to another family member. A bit more complex use case can be a seller provides loans to customers who intend to purchase a house. More complex use cases can be Distributed Autonomous Government (DAG) where settlers code and enforce their own governmental services.

The researchers reviewed the top 30 blockchains, around 17 blockchains only were found to have established smart contract support. However, out of those, only 10 are not forks of each other. Table 1 shows some of the most popular smart contract blockchain platforms sorted in descending order based on market capital.

With a myriad of smart contract platforms available, a question arises; How do these platforms compare with one another? Comparing various smart contract platforms has been attempted many times. Blockgeeks [10] attempted to compare some of smart contract platforms, namely, Ethereum, EOS, Stellar, Cardano, Neo, and Hyperledger Fabric. According to the authors, smart contracts should have the following desirable features: determinism, termination, and isolation. The authors then compared smart contract environments (Virtual machines vs Dockers) to determine the environment that best caters for smart contracts. The authors have found that the Virtual Machines provide better ecosystems for smart contracts. The authors also

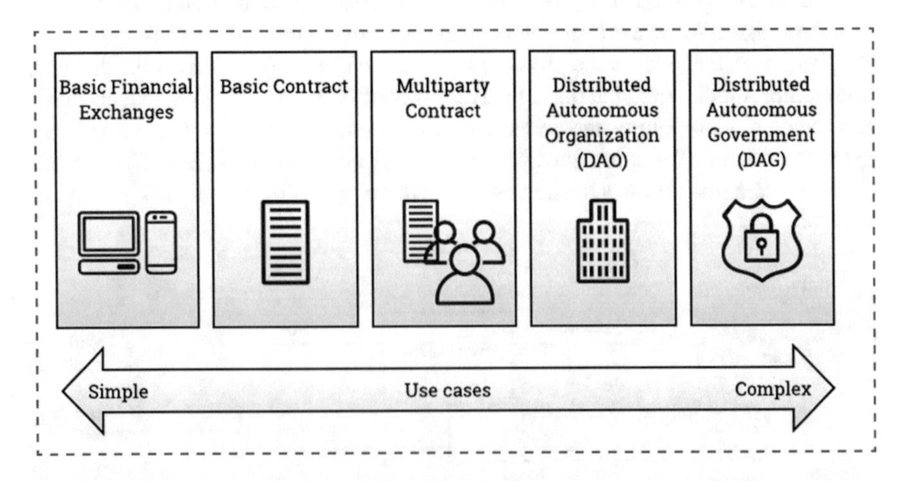

Fig. 4 Smart contracts can range from simple basic currency transfer contracts to Distributed application, organizations, and societies [36]

Table 1 Some of the most popular Smart contract platforms

Platform name	Market capital	Date released	Cryptocurrency (denotation)
Bitcoin	$72,201,465,881	January 2009	Bitcoin (BTC)
Ethereum	$14,848,712,547	July 2015	Ether (ETH)
EOSIO	$3,913,070,561	January 2018	EOS (EOS)
Stellar	$2,064,638,315	July 2014	Lumens (XLM)
Cardano	$1,781,215,623	September 2017	ADA (ADA)
TRON	$1,551,622,409	May 2018	TRON (TRX)
Ontology	$614,045,134	March 2018	Ontology (ONT)
NEO	$609,054,336	June 2017	NEO (NEO)
Tezos	$570,618,595	September 2018	Tezos (XTZ)
Waves	$279,346,879	End of 2016	Waves (WAVES)

presented the advantages and disadvantages of each of the platforms compared. Their comparison was based on transaction cost and speed, scalability, and development language. The authors argue that Ethereum might be the de facto platform, and is truly decentralized, but EOS and Stellar might be a better choice for scalability, and low fees. The authors concluded that each SC platform has a unique feature that makes it desirable. The authors provided some important arguments, but followed no formal evaluation benchmarks.

Other attempts have been made to evaluate smart contract platforms such as Mulders [11], Max [12], Chiu [13], and others who briefly discussed a number of prominent smart contract platforms that include: Ethereum, NYM, Hyperledger Fabric, NEO, Cardano, and Stellar. These platforms were compared based on their main attraction features and other criteria such as median transaction confirmation time, transaction price, features, and security features. All of the above non-formal research efforts have compared the platforms based on random non-deterministic comparison criteria, followed no formal comparison methodology, and parts of the evaluation were done based on personal points of view without empirical assessment. The source data used in the comparison was also not verified for accuracy.

Formal research in smart contracts on the other hand, is thriving. According to Macrinici et al. [14], smart contract research in 2017 has almost tripled in number since the year 2014. The researchers have also found that most of the research is related to the blockchain security, privacy and scalability, and with regards to smart contracts, most research is directed towards the programmability of smart contracts. Evaluating the merits/pitfalls of smart contracts in formal research specifically can be considered in its infancy stage. Delmolino et al. [15] in their attempt to teach smart contract programming lab in the University of Maryland, have faced and documented the pitfalls they faced. This is a form of evaluation of the smart contract programming and can be categorized into two main categories: User pitfalls (such as logical errors, and use of cryptography), and development platform specific pitfalls (such as Ethereum specific bugs).

Bartoletti and Pompianu [16] have attempted to empirically evaluate smart contracts implemented on the Bitcoin and Ethereum blockchains. In their research, the authors collected a sample of 834 smart contracts deployed to the Ethereum and bitcoin blockchains, then manually attempted to analyze and evaluate them based on the following criteria: application domain, and design patterns. The authors have also presented comparison of some of the prominent smart contract platforms based on their key features, similarities, and differences derived from a published. The paper was mainly focused on only the bitcoin and Ethereum blockchains in the comparison. The authors have found that the majority of contracts deployed in Ethereum were either financial, game, or unclassified contracts. Bitcoin blockchain had the majority of Notary followed by financial transactions. Overall, the majority of transactions executed on those smart contracts were of the financial nature, then followed by the Notary transactions. Concerning the design patterns, the authors have found that token, authorization, time constraint, and termination are the programming patterns used the most among the sample of smart contracts evaluated. The research paper was detailed and empirical in nature, however the research was only limited to two evaluation criteria.

Presently, non-formal research (non-refereed articles published over the internet) contained comparisons of many of the front running smart contract platforms; however, none of the comparisons was based on standard framework to evaluate smart contracts and could be easily considered as non-impartial or even biased comparison. Meanwhile, most of the formal research was limited to a number of criteria and has studied the Bitcoin, or Ethereum blockchains alone. Very limited research is performed on other blockchains.

In our work, and based on previous work, we intend to adapt an established evaluation framework to the smart contract landscape by capturing the criteria that evaluate the current state of smart contracts such as usage, maturity, security and privacy, programming patterns/languages, and performance as shown in Table 2, and creating a newly adapted evaluation framework. The defined framework will then be used to empirically evaluate smart contracts in some of the most dominant smart contract platforms available; such as Ethereum, Stellar, and EOS.

2 Smart Contract Evaluation Framework

Many references were reviewed with the goal of collecting relevant information related to evaluation tools. Either of the following two paths has to be followed; building the framework from scratch or adapting an existing framework. Adapting an existing well-established evaluation framework was selected for efficiency. The evaluation framework of Heitkötter et al. [17] who developed a framework to evaluate mobile development platforms. The framework was selected for its comprehensibility and criteria relevance to software development platforms. The original framework contained 14 criteria in total split into two parts: 7 Infrastructure related criteria, and 7

Table 2 Summary of previous work

	Compared platforms	Compared criteria
Blockgeeks [10]	Ethereum EOS Stellar Cardano Neo Hyperledger Fabric	SC environment Main advantages of each platform Transaction cost Transaction speed Scalability Development language
Mulders [11]	Ethereum NYM Hyperledger Fabric Stellar	Main blockchain attraction features Median transaction confirmation time Transaction price Security
Max [12]	Ethereum Neo Lisk EOS Cardano	Advantages of each platform Development language Development environment Market capitalization
Chiu [13]	Ethereum NEO Cardano ADA	Consensus Current and future Transactions Per Second (TPS) Eco system/project development
Bartoletti and Pompianu [16]	Bitcoin Ethereum	Application domain Design patterns

development related criteria. The criteria in the original framework were well developed and provided solid ground for building the smart contract evaluation framework. To adapt the framework to smart contracts, the researchers considered the criteria included in previous works detailed in Table 2. The resulting framework is composed of 4 Infrastructure related criteria, and 7 development related criteria. The framework was finally evaluated for its reliability and validity. In the subsequent section, we provide more details about the evaluation framework and its related criteria.

2.1 Evaluation Framework Infrastructure-Related Criteria

A summary of all the infrastructure related criteria can be found in Table 3.

I-1 License and Costs

Smart contracts are processed in the form of transactions on the blockchain. To prevent Denial of Service (DoS) attacks on the network, spam, and to decide the priority of transactions to be mined on the blockchain from the pool of pending transactions [18, 19], fees are usually imposed on each transaction. To ensure fairness, this fee is usually proportional to the content. This criterion reviews licenses associated with use of smart contracts and evaluates the costs based on the following two items:

Table 3 Summary of Infrastructure related evaluation criteria

Serial	Criterion	Description
I-1	License and costs	Total costs associated with deploying a plain contract on the blockchain
I-2	Sustainability	Investigates the sustainability and future proofing measures of the smart contract and the hosting blockchain
I-3	Speed and performance	Performance and processing speeds of the smart contract. Such as, total time taken to execute a transaction, and the average block time
I-4	User experience	Measure the level of user satisfaction

1. Measuring the average fee paid for a sample ICO contract transactions on the blockchain.
2. Average transaction costs on the blockchain in the last 200 blocks. This was selected based on two factors: blockchains are of varying chain lengths, and blockchain transaction costs may vary with time. Considering the costs of the last 200 blocks would be a suitable measure.

I-2 Sustainability

Software is in an ever-changing state. Bugs are patched, new features are implemented, and updates are added incessantly. Will the smart contract remain functional in the future time to come? This criterion examines the measures emplaced ensuring that the SC doesn't suddenly brake or stop working whenever an update or new compiled version is released on the smart contract platform. This criterion also examines the predicted state of sustainability to the blockchain itself.

In summary, this criterion examines the sustainability of:

1. The smart contract.
2. The hosting blockchain. We will use the following items to measure the sustainability of the hosting blockchain:
    - Number of DApps and users overtime
    - Number of newly acquired addresses overtime
    - Number of transactions executed on the blockchain overtime.

I-3 Speed and Performance

Blockchains have varying numbers of users with varying usage patterns that affect the processing speeds of transactions on the blockchain. This criterion evaluates the speed it takes to publish the SC to be available on the blockchain, and the time it takes for the contract to execute. For evaluation, the sub criteria are:

Average transaction execution time: The average time taken from the execution of a smart contract transaction to the time it is mined and available on the blockchain. It can be a transaction to publish a smart contract or a transaction that executes the smart contract, such as a transaction that transfers funds once the contract conditions are met.

Average block time: It is the time taken to mine a new block on the blockchain. For example, in the bitcoin blockchain, it is expected that the average block time is 10 min. Blockchains use varying difficulty algorithms to control the generation of new blocks on the network. Block times are enforced by the difficulty level of this algorithm. With an increased block time, the transaction throughput time increases, therefore, the lower the block time the better the smart contract performance will appear.

Number of Transactions processed Per Second (TPS): The number of transactions the blockchain can process in a second at a given time. TPS is calculated using the following formula:

$$\mathrm{TPS} = \mathrm{Nb} * \mathrm{Ntpb}/\mathrm{Tb}$$

where Nb is number of blocks or ledgers, Ntpb is number of transactions per block and Tb is the time to generate the same blocks in seconds.

I-4 User Experience (UX)

This criterion evaluates the users's experience of these smart contracts in terms of the comfort, acceptability of use of the smart contracts, and the overall satisfaction level. This criterion can be measured using any of the traditional data collection tools such as questionnaire or interviews.

2.2 *Evaluation Framework Development-Related Criteria*

A summary of all the development related criteria can be found in Table 4.

D-1 SC Development Environment

Smart contracts are at the core software code that is developed using software development tools. This criterion evaluates the features and robustness of the SC development environment itself by measuring the memory and disk usage of development framework and ease of installation. This criterion also explores the following features:

- Development environment specifications:
  - Type of development environment: Software Development Kits (SDKs), desktop-based, web-based development environment, etc. and the availability of debuggers and compilers.
  - Memory and disk usage of development environment.

Table 4 Summary of development related evaluation criteria

Serial	Criterion	Description
D-1	SC development environment	Development environment ecosystem specifications Type of development environments available Network, memory and disk usage of development framework Availability or lack of the following features: Test environment availability Exception handling, logging, and event logs
D-2	SC development environment legal considerations	Evaluate the availability or lack of the following legal considerations: Time constraints: contract starting and expiration dates and timed variables if any Private variables/secret fields, private contracts Authorization and multi signature capabilities Evaluate SC automated validity deterministic verification
D-3	SC security features	Investigate the availability of smart contract security functions and features
D-4	Ease of development	Wealthy documentation Active community Development language
D-5	Flexibility and modifiability	Investigate the modification options
D-6	Code reusability	Investigate code reusability options
D-7	Speed and cost of development	Environment setup speed, and development speed

- Availability or lack of the following features:
  - Test environments
  - Exception handling, logging, and tamper proof event logs: these are necessary in case of disputes and incidents that require conflict resolution.

D-2 SC Development Environment Legal and Ethical Considerations

This criterion evaluates the features of the SC development framework/language that makes it capable of developing SCs that compete with traditional legal counterparts. This criterion also examines the SC validity respects. This criterion considers the following:

- Availability or lack of the following legal considerations:
  - Time constraints: starting and expiration dates and timed variables if any

  - Private variables/secret fields, private contracts
  - Authorization and multi signature capabilities.

- SC automated validity verification: existence of some method or tool that deterministically compiles/verifies the smart contract code to ensure the following:

  - Guarantee that the contract code is doing what was agreed upon
  - Guarantee that the contract code does not do anything extra
  - Guarantee that the contract code does not change behavior overtime.

D-3 SC Security and Integrity Features

SCs require strict security measures to be in place. This criterion investigates the availability of SC security measures provided by the programming language and security testing tools. For example, does the smart contract platform provide encryption functions? Does the blockchain provide measures to check/detect orphan addresses during runtime to prevent money loss [20]?

D-4 Ease of Development

This criterion sums up the quality of documentation and the learning-curve. A well-documented SC platform with peer discussion forums, and a wealthy documentation with code snippets to back it up, with a common development language may present an easier and shorter learning curve and knowledge transfer between developers than other counterparts. This also provides easier training of new developers on the SCs. This evaluation criteria includes evaluation of the following:

- Wealthy Documentation
- Active community
- SC Development Language used.

D-5 Flexibility and Modifiability

Once a SC is written to the blockchain, it becomes Immutable. This is a cause for concern if the SC contains bugs, or if contract conditions were altered for any reason. This criterion evaluates the flexibility and ease of modifying an existing SC. This criterion can be measured using user survey response.

D-6 Code Reusability

SCs are simply just software code. This criterion determines the reusability of source code or parts of it from one SC to another. This could be very valuable in cutting down development cost/effort.

D-7 Speed and Cost of Development (Efficiency)

This criterion evaluates the speed of the development process and factors that hinder or speedup a straightforward development. Three items to consider:

1. Environment setup speed: the time required to setup a development environment.

2. Development speed: The time the developer starts developing of the SC, to the time the code is ready.
3. Development language learning speed: The time is takes to learn the development language.

3 Evaluation Framework Quality Test

Researchers rely on two different dimensions to evaluate their measuring tools; Reliability and validity [21, 22]. To ensure that the framework is consistent and that it measures what it is intended to, the same dimensions were considered:

Reliability
It "refers to consistency or stability of measurement" [23]. When judging the reliability of the measure, researchers can consider one or more of the following types: Test-Retest Reliability, Internal Consistency, and Interrater Reliability. Test-Retest Reliability: refers to the extent which the measurement/score of a criterion remains consistent overtime. Internal Consistency: refers to consistency of responses/scores of multiple items measuring the same underlying construct. Interrater Reliability: refers to the extent which observers are consistent in their judgement of a measure [22].

In this research, we have relied on Test-Retest reliability testing to judge the reliability of the framework. The instrument was applied twice to the field of study. The first application was on (12 February 2019). The instrument was reapplied 17 days later, (1 March 2019). We tried to ensure that the application conditions remained as similar as possible to the first application.

Validity
It "refers to the suitability or meaningfulness of the measurement" [23]. To judge the validity of the framework, as well as to avail from the experience of the raters in the field of study, the researchers used the qualitative validity method of evaluation. 100% of the framework raters agreed explicitly or implicitly that the evaluation framework is valid.

4 Comparative Study: Ethereum, EOSIO, and Stellar Smart Contracts

In this section we present the results received after applying the evaluation framework on the following blockchains: Ethereum, EOSIO, and Stellar.

4.1 *Tools and Experiments Used to Apply the Evaluation Framework*

This subsection will present the experiments and tools used under the framework to obtain results across various smart contract platforms.

License and Costs

To reach an evaluation of smart contract costs of the sample blockchains, licensing and cost was approached from the following perspectives:

- Measuring the average fee paid for a sample ICO contract transactions on the blockchain. Ethereum transaction calculator[1] was used to estimate the cost and processing time of the transactions on the main net. Metamask[2] was used to obtain the predicted values from the Ropsten Testnet. Ethereum smart contract development was done on Ethereum Remix Integrated Development Environment (IDE). Actual values are obtained from the Ropsten.etherescan website [24] after deployment. Stellar Laboratory was used to deploy and obtain values from the Stellar test net. EOS BEOSIN[3] along with Scatter[4] were used to deploy and obtain values from the EOS Jungle testnet.
- Average transaction costs on the blockchain in the last 200 blocks. This was manually calculated from data obtained at each blockchain's official block explorer at the same period of time. Ethereum block data was collected from etherscan, the official block explorer. Stellar data was collected from the Stellar official dashboard. EOS data was more complex to obtain as data was available in many explorers such as eostracker.io, eosnetworkmonitor.io, eospark.com, eosflare.io, bloks.io and more developed and maintained by various companies and official EOS block producers. Data was collected from at least three sources and the results were compared to ensure accuracy. The above referenced block explorers will be referred to as block explorers or blockchain explorers hereinafter.

Sustainability

Since it was not easy to measure the smart contract sustainability, the researchers tried to view the issue from the opposite perspective. With smart contracts deployed on the blockchain being immutable, they will remain tamper proof and operational in the coming time, but what could stop a fully working smart contract from running in the future? The answer to this question came as follows:

---

[1]Ethereum transaction calculator. Available at https://ethgasstation.info/calculatorTxV.php.

[2]Metamask is an application that enables the user to communicate with the Ethereum blockchain without the need to deploy a full node. Available at https://metamask.io/.

[3]BEOSIN is an online IDE for EOS smart contract development, testing, and deployment. Available at https://beosin.com/BEOSIN-IDE/index.html#/.

[4]Scatter is a multi-blockchain platform (EOS, Ethereum, Tron) wallet and asset management software. Available at https://get-scatter.com/.

1. The smart contract would no longer be usable if the underlying low-level opcodes or byte programs were stopped from executing on the Docker or Virtual Machine. To study this item, the changes in opcodes and smart contract code were investigated and the official community forum and the official documentation were reviewed for evidence of opcode changes or bytecode deprecation.
2. The smart contract would no longer be usable if the blockchain platform such as Ethereum EVM for example or Stellar blockchain no longer supported smart contracts. This is an unlikely scenario because these platforms have specialized and differentiated themselves from other platforms by supporting the smart contracts. Their main appeal is the smart contract support, hence the rule out of this possibility.
3. The smart contracts would cease to exist if the underlying blockchain platform itself was stopped from use by the community. But it is difficult to predict if the application will remain in use in a future time to come. Instead, the researchers tried to observe how the blockchain has been growing so far to deduce future growth trajectories to measure the sustainability of the blockchain itself. The following items were used to measure the sustainability of the hosting blockchain:

   - Number of DApps over time. This data was acquired from DApp Radar,[5] and from Stateofthedapps [25].
   - Number of newly acquired addresses overtime. This data was obtained from the various blockchain explorers.

Speed and Performance
The following items were measured:

- **Average transaction execution time**: The average time taken from the execution of a smart contract transaction to the time it is available on the blockchain is measured using the etherscan.io for Ethereum, Stellar's transaction explorer for Stellar smart contract transactions, and jungletestnet.io[6] for EOS testnet transactions.
- **Average block time**: This value can be obtained directly from block explorers.
- **Number of Transactions processed Per Second (TPS)**: Having the average block time, the number of transactions per second can be calculated using the following equation:

$$\mathrm{TPS} = \mathrm{Nb} * \mathrm{Ntpb}/\mathrm{Tb}$$

where Nb is number of blocks or ledgers, Ntpb is number of transactions per block and Tb is the time to generate the same blocks in seconds.

User Experience
The measurement of the satisfaction level was initially planned to be measured via

[5]DApp Radar. Available at https://dappradar.com/.

[6]jungletestnet.io. Available at https://monitor.jungletestnet.io/#home.

Table 5 Interviewee demographics

Years of experience	Main field of expertise
3 or more	Ethereum and Bitcoin
2 or more	Ethereum
2 or more	Ethereum
Around a year	Ethereum
1 or more	EOS and other blockchains
Around a year	Stellar

online survey. The survey was announced in Bitcoin talk forum, sent to individuals via email, and announced on Reddit social media website. Survey was opened for a period of over two months, but only scarce responses were received by the community. The data collection method was then switched to interviews and approached a select set of individuals for interviews based on the length of time spent developing smart contracts. The longer the duration the more expertise of the individual Interview questions were those included in the survey. The sample of interviewees included 6 individuals from the GCC region as well as the USA and Canada. The sample were blockchain developers or blockchain based company owners/managers. Table 5 presents the sample demographics. The user satisfaction level for each criterion was qualitatively devised from the user responses. The overall satisfaction level was also devised from the user feedback.

SC Development Environment

Initial inspection of the development ecosystem from the community as well as official developer resources were used. This research has resulted in the finding of various development approaches and tools.

SC Development Environment Legal Considerations

The official development language documentations of Solidity, EOS github repository, and Stellar Laboratory to collect evidence of availability of legal considerations or lack thereof were reviewed. The researchers relied on the responses of the interviewed sample with regards to the SC validity verification.

SC Security and Integrity Features

The availability or lack of security classes and functions were investigated, in addition to the best practices and known bugs in the official development language documentation and blockchain github.

Ease of Development

This criterion evaluates the following three items: Wealth of documentation, Active community, and Development language:

Wealth of Documentation. This item was measured by researching the available official documentation, and the quality of the documentation itself based on the following items: The frequency of update, searchability, quality of the documentation in terms of quality of language as well as formatting, availability or lack of examples, in addition to the community documentation and online tutorials.

Active community. Researchers tried to measure the community activity by considering the concentration of users in the official pages on the major blockchain social media networks.

Development Language. The development language options available that can be used to code the smart contract were explored. Details were gathered from the official blockchain documentation.

Code Reusability

To examine the source code reusability level from one smart contract to another, the responses of the interviewed sample was used as a measure.

Flexibility and Modifiability

As software is man-made, and constantly subject to modifications, corrections, and optimizations, modifiability becomes a necessity in smart contracts. To measure this, the responses of the interviewed sample as well as research were collected to identify modification options the community are devising to work around the issue of smart contract immutability.

Speed and Cost of Development

Experimentation has been conducted using the development tools/languages to measure the speed and cost of development details.

4.2 *Evaluation Framework Results*

This subsection describes the obtained results of applying the evaluation framework on the three smart contract platforms: Ethereum, EOSIO, and Stellar.

License and Costs

From Table 6, it is evident that Ethereum is the most expensive of the three blockchains when looking at the average transaction costs. Chen et al. [26] has designed a gas reducer that detects under-optimizations in the contract code to be replaced by more optimized and efficient code, therefore reducing the overall transaction price. This only further proves that smart contract cost is a community concern. As the number of Ethereum main net users grows, the prices are only expected to

Table 6 Average transaction costs in Ethereum, Stellar, and EOS blockchains

	Ethereum	Stellar	EOS
Average transaction costs on the main net (USD)	0.72	0.00000141	No fees
Average transaction costs of the last 200 blocks (USD)	0.12	0.00005300	NA

rise. These results are in line with the observations of Macrinici et al. [14] who stated that the Ethereum gas prices are getting more expensive. Nonetheless, average transaction costs of smart contracts on the blockchain are negligible and much lower than the costs associated with traditional contracts.

Sustainability

Overall, smart contracts are said to be more sustainable than traditional counterparts, because they reduce the execution time, the middleman, and costs. Stellar smart contracts seem to be the least sustainable of the three. As for the sustainability of the evaluated blockchains themselves, it can be said that they are all equally sustainable. It was also found that the blockchains and smart contracts seem to have growing adoption. With time, they are expected to dominate a large part of our lives. This finding is in line with the findings of Avan-Nomayo [27]. In his article, the author believes that cryptocurrencies are not yet the mainstream. He states however, that cryptocurrencies seem to be following the same growth the internet followed in the early 1990s and 2000s and if the cryptocurrencies continue to follow the same trends, then they will most likely dominate the future.

Speed and Performance

According to the data in Table 7, EOS can be said to have the best performance among the three blockchains. It has the most consistent block time among the three blockchains generating a new block in exactly 0.5 s block time, and an impressive 74.8 transactions processed per block.

User Experience

Overall, the Stellar smart contract platform can be considered as the most satisfactory blockchain among the three based on user response. The user satisfaction levels were higher than those of the other two blockchains having the relatively high transaction speeds, the UI based laboratory, and the low transaction costs as the most appealing factors. The Ethereum blockchain was considered to have unrivaled community support, while the EOS blockchain WASM feature was its most appealing factor.

SC Development Environment

Ethereum can be said to have the best DApp development ecosystem because the number of tools and their robustness as compared to the other platforms. This could be attributed to the age of the blockchain, as with time, tools and supporting modules are sure to mature. The blockchain smart contract eco system is

Table 7 Average speed and performance metrics of Ethereum, Stellar, and EOS blockchains

	Ethereum	Stellar	EOS
Average transaction execution time (sec)	48	5.64	0.5
Average block time (sec)	19.6	5–6	0.5
Number of transactions processed per second (TPS)	19.4	3.3	74.84

constantly changing, and it is possible that all these requirements will see further improvements in the coming future.

SC Development Environment Legal Considerations

Concerning availability or lack of the following legal considerations: time constraints, private variables/secret fields, private contracts, and authorization and multi signature capabilities, Stellar has the basic legal considerations covered, but the other blockchains do offer more flexibility when it comes to creating complex smart contracts. As to the SC automated validity deterministic verification, no formal validation is currently available unless with manual use of hash functions. This is a concern in both EOS and Ethereum blockchain smart contracts. This is not much of a concern in Stellar smart contracts due to the transaction builder used to build smart contracts which enables the user to self-build easily the contract rather than rely on a third party.

SC Security and Integrity Features

Security vulnerabilities are not foreign to any development platform. Blockchain smart contract platforms are not an exception. However, each of the smart contract development platforms tries to maintain a security best practices, Stellar seems to have the least surface of vulnerabilities. Having an easy to use transaction builder, with the pre-developed operations, users are able to create smart contracts that are quick to develop and are generally more secure than hard-coded smart contracts. The EOS platform seems to have the most skepticism from the community towards the level of security of WASM contracts on EOS. Perez and Livshits [28] have found that although smart contract vulnerabilities might be important to study, however they are subject to great exaggeration.

Ease of Development—Wealth of Documentation. Documentation is of prime importance because it enables new developers into the field, increases adoption rate, knowledge transfer, and as a result, development momentum. Despite this importance, we find that the documentation in all the blockchains is somewhat neglected. They were found to be hard to follow, and can be even confusing at times for new entrants into the domain. Ethereum has been around the longest, and has quite extensive documentation that can be considered the best of the three. Stellar has been around for a while as well, yet the documentation is not as extensive as one expects. EOS documentation is quite summarized and can be exceptionally hard to follow having limited experience in the field. Overall, the documentation needs better development. Especially the documentation to new entrants to the smart contract domain, different ways to approach development, the underlying blockchain inner workings, security functions, mechanisms, and caveats, deployment options etc.

Ease of Development—Active Community. Overall, Ethereum has the most active community, followed by EOS. Stellar has the least active community among the three blockchains as seen from Fig. 5. This could be attributed to multiple factors such as the age of the blockchain, as well as the popularity among the community users, and adoption rates.

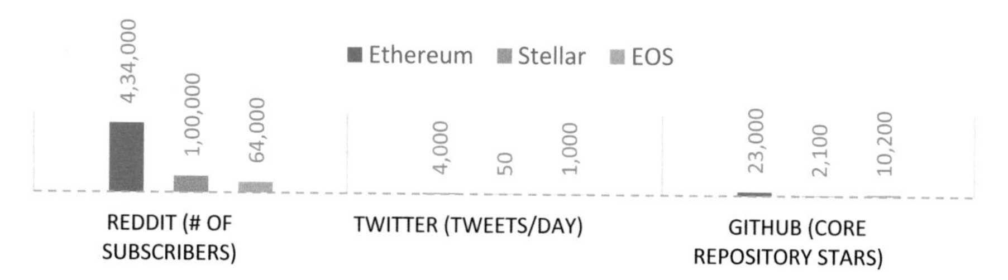

Fig. 5 Blockchain Community rates among the popular social networks

Table 8 Blockchain main development languages

	Ethereum	EOS	Stellar
Main development language(s)	Solidity	Any language that compiles to WASM is supported	No language

Ease of Development—Development language. As shown in Table 8, Stellar has no development language because it uses a UI transaction builder. EOS is considered to have the better development language when compared to Ethereum. Supporting WASM provides EOS smart contracts flexibility in the development language in that it enables developers to develop smart contracts in the development language of their selection as most development languages compile down to WASM.

Code Reusability

Code reuse on the one hand can mean great risks by reusing vulnerable software code, and propagating software bugs among smart contracts. However, on the other hand, reuse of compartmentalized code and code modularization could also mean that secure and tested bits of code, can easily be reused to save time and effort and mitigate attacks. Among the three blockchains, we found that code reusability has the least significance in the Stellar blockchain. It was also found that code reusability is facile among the Ethereum users and more common as described in Table 9. Many smart contracts deployed of the Ethereum network have more than 100 lines of code. 100% of the smart contract developers have had a reuse of either their own or another's smart contract code. Half to most of the code is often reused by the Ethereum smart contract developers. There are also many resources, smart contract

Table 9 Code reusability

	Ethereum	EOS	Stellar
Developer reuse	Reuse whenever possible	Reuse whenever possible	NA
Templates and ready code	Available in abundance	Some resources available	Not available

Table 10 Smart contract modifiability across study blockchains

	Ethereum	EOS	Stellar
Contract modifiability	Persistent	Versionable	Persistent

templates, and pre-made contracts available in Solidity for the Ethereum community to easily reuse (i.e. OpenZeppelin).

Flexibility and Modifiability

EOS seems to be the single platform that supports versioning and modification of smart contracts. This is attributed to the nature of contracts being stored on the deploying node. Once smart contracts are deployed to the blockchain itself, they are no longer modifiable. Moreover, contracts are not only immutable, but also replicated. So, once a smart contract is published to the blockchain, it is not only non-modifiable, it is also persistently stored as is in thousands of nodes across the globe.

Although neither Ethereum nor Stellar platforms provide any modification capabilities or standards (Table 10), Ethereum can be considered the closest to having some sort of option to destroy an existing smart contract. This demonstrates the need for a set of standards or tools that assist in the alteration of modification of deployed smart contracts. This finding is in line with the observations done by Marino and Juels [29]. The authors saw the need to develop a set of standards that govern the alteration and undoing of smart contracts.

Speed and Cost of Development

Stellar seems to be the easiest platform to setup and use. Ethereum and EOS however have convoluted development environment that requires a large amount of time to setup the development environment. Of all three, EOS takes the most amount of time to setup the development environment, and Ethereum seems to take the most amount of time to learn and develop a new language as shown in Table 11. Overall, investment in research is still a mandatory requirement from the developers to get started.

Table 11 Environment setup speeds of Ethereum, EOS, and Stellar

	Ethereum	EOS	Stellar
Environment setup speed	A day to a couple of days	A couple of days to a week	None
Development speed	120 h to a few weeks	100 h to a couple of weeks	120–200 h
Development language learning speed	A few weeks	Negligible	None

5 Smart Contract Recommendations

When it comes to smart contracts platform selection, there exist plenty of platforms to choose from. Normally, the selection of the platform would be based on stakes owned or platform personal preference. However, if those were not players in the decision-making, an informed decision to select the optimal platform should be based on sound evaluation of requirements, and development considerations. Each platform has its own niche, advantages, and disadvantages. Hence, the platform of selection should be done on a case-to-case basis. When it comes to DApp development for example, from Table 12, it can be seen that minor differentiation can determine the blockchain most suitable for the use case. If the response time is insignificant, and the data is highly critical, then a blockchain with high validity such as Ethereum can be used.

If the response time is critical, as well as the data, such as critical applications, then a highly scalable blockchain such as EOS can be used. If the data is not critical, but the response time is of the essence, then a highly responsive blockchain, or traditional applications and databases can be used. If the data is not critical, and the response time is not urgent, then any blockchain of selection or any traditional application model can be used. A smart contract requires strenuous effort to develop and ensure that it guarantees to perform what it was intended to do, therefore a need arises for ethical and strong software development standards and stringent testing exercises to ensure high quality standard smart contracts and DApps. Developers are recommended to use the established software engineering practices and apply software quality tests since smart contracts are pieces of code at the end of the day. This will raise user trust and as such, will result in higher adoption rates. Development of smart contracts as depicted in Fig. 6, is recommended to follow the following workflow: Development and deployment on a local test net. Testing and debugging iterations follow until the smart contract is ready to be deployed to the public test net. The public test net can provide more lifelike scenarios that might be faced on the main net. Public test net

Table 12 Usability matrix

	High criticality data	Low criticality data
High response time	High scalability blockchain	High scalability blockchain\Traditional apps
Low response time	Ethereum blockchain	Any selected blockchain\Traditional apps

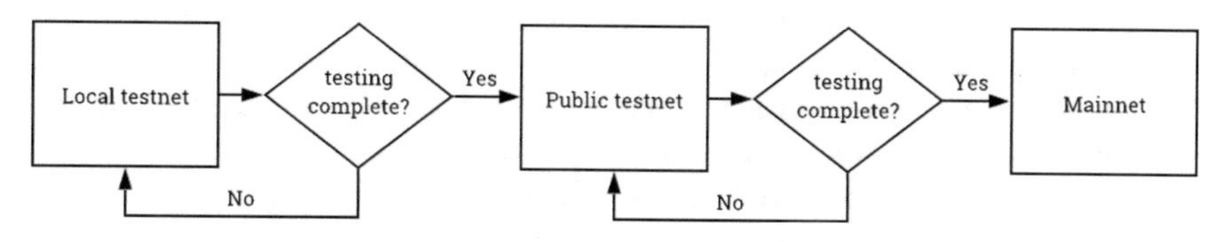

Fig. 6 Recommended smart contract development workflow

testing iterations follow until the smart contract is bug free, to finally reach the phase where deployment to the main net is possible.

Smart contracts are highly susceptible to collusion and therefore hashing practices are required to counter deception and ensure that the agreed contract is indeed the contact that was deployed to the blockchain. A smart contract developer cannot develop a smart contract, publish it on the blockchain in its current capacity, then forget about it and rely on it completely to do what it was programmed to do. The developer should be aware of the need to be in a constant state of research and awareness, continuously remain up to date with proposals and changes to be able to cope and update the smart contract during its lifetime. Low level programming code might change, and become no longer usable, hard forks could occur that forces the developer to migrate the contract to the new blockchain, and loopholes in the code require constant maintenance considering the critical nature of smart contracts. In addition, smart contract validity is of prime importance. Unfortunately, currently there is no blockchain consensus verified validation of smart contracts, therefore, to build trust, the smart contract stakeholder has to review the source code manually, and then compile it and generate hash values of the compiled code to ensure that the source code does what it is intended to do, and does not change.

6 Research Topics in Smart Contracts

In conducting this research, many lessons were learned. The blockchain smart contract domain is vast and wide, and without having to come to realizations, has numerous possibilities for further research. The following are some of the areas that require future research:

- **Differences between traditional and blockchain DApps**. More studies are required that compare between blockchain DApps that use the smart contracts, and blockchains in the build and traditional DApps developed using traditional development methods and DBs. This is to ascertain the actual benefits the blockchain DApps deliver as compared to the traditional software DApps.
- **Nonexpert user perspective**. Smart contracts are at the core software programs. Like in all software programs, software quality is of the essence. With smart contracts being the handlers of financial transactions on the blockchain, they become a major security risk and the software quality becomes much more important. As seen, creating smart contracts in the present requires technical knowledge, software development background, and investment in research and large effort trial and error to learn and develop trustworthy, vulnerability free smart contracts. The closest form of contracts that is nonprofessional friendly is the Stellar platform, and even that requires some investment. Even if the end user decided to hire a professional for their smart contract, trust in the developer/auditor is still required when it comes to smart contracts. To reduce this trust dependency, Knecht and

Stiller [30] have proposed a smart contract platform that aids in creation, deployment, and management of smart contracts. Overall, more research is required to create more user-friendly smart contracts. Bragagnolo et al. [31] have created a smart contract visualizing software that enables developers to debug the state of a contract after it is deployed to the network. Mavridou and Laszka [32] have created a web application for designing and generating secure Solidity smart contracts. This application is a GUI based application with a set of design patterns and plugins that enabled developers to easily enhance functionality and security of their smart contracts. The application is open source and has about 184 commits. These efforts only stem from the need to have user-friendly interfaces for the nonprofessional to easily use and create trustworthy smart contracts.

– **Cross platform interactions**. As we have in the real world, a contract can have many terms, clauses that require different currencies. But unlike the real world, blockchain cryptocurrency smart contracts do not interact with each other. An Ethereum smart contract cannot have terms that communicate or control funds in an EOS or Stellar blockchain for example. It would be a very beneficial possibility if blockchains were able to interact with each other, without having to create/obtain new currency only to use a smart contract because it does match with the currency at hand. Possibly due to unique characteristics of each platform, there are rarely any tools that are extendable to other platforms.
– **Software engineering and development perspective**. Software engineering is concerned with the application of various principles to ensure quality software systems that best satisfy the requirements of the users. With the SC development being a novel occupation, it is highly favorable that the SC development is disciplined with tailored software development methodology. Numerus incidents have been reported on the Ethereum blockchain that cause concern to the quality of smart contract software developed. Such incidents include the infamous DAO Hack, which took place in 2016 where hackers exploited a vulnerability in the smart contract code and stole 3.6 million Ether, which is around $60 million at the time [33]. Another hack just as grave took place in 2017 where an unknown hacker stole around $32 million by exploiting the delegate call and fallback functions in Parity's Multi-signature Wallets. Well known software development practices such as use cases, code reviews, testing, and correctness proofs can be applied to smart contracts just as they apply to any other form of software. But blockchain smart contracts require their own considerations, standards, and tools. From this context, and looking at the present state of smart contracts, it can be said that smart contract development in the present lacks formal software engineering customized models. For example, issues that require modifiability are commonplace, and with not enough knowledge or experience, and lack of careful planning, a developer could run into issues that invalidate the entire smart contract. Therefore, standard practices and software engineering processes are required to raise the smart contract software quality. Porru et al. [34] and Destefanis et al. [35] have confirmed that there is a need for a formal discipline that defines best practices and formal methodologies concerning the development of smart contracts on blockchains. Another hindrance is that the development tools, languages, and

technology have a steep learning curve. Unification of smart contract development technologies such as development language and development platform or at least unified standards could contribute to decreasing this learning curve.

- **Various smart contract blockchains**. A large volume of research is found to be based on the Ethereum network. With the growth in market capital and growing adoption rates of other blockchains, such as Ripple, Stellar, and EOS, research on other blockchains is also required. If available, they would aid in leveraging the blockchain technology and encourage further advancements.
- **Smart contract technical development**. There are several difficulties with the smart contracts development that require further research such as random variables, modifiability options, and automatic testers, bug detectors, and inspection tools to identify values of runtime variables could be beneficial to developers. A lot of skepticism was felt from the EOS community with regards to the security of WASM integration with the blockchain. This could also be a very important topic for future research. Performance wise, more research is required on congestion and stress tests, possible ways to control or reduce the growing transaction fees in Ethereum, and RAM prices in EOS. More research on optimization algorithms and deep learning and the ways they can also be integrated with smart contract to increase the quality of smart contracts.

7 Conclusions

The true value of smart contracts emerges when combined with already established technologies forming a new breed of applications; DApps. Smart contracts are hardened once embedded into the blockchain. They become non-modifiable and require strenuous effort for anyone to breach. This provides a guarantee that the terms of the contract will be met, and the funds will be transferred in a timely manner without any violation to the original agreement. They promise to provide transparency in their transactions, trust in a distributed trustless environment, and an economic boost in a non-centralized world. Existence of various blockchain platforms, introduced various ways to develop and produce smart contracts. The development procedure and the quality of smart contract using various platforms haven't been measured based on a well-defined framework or criteria. In this study, an empirical evaluation for the most prominent smart contract platforms in the market based on a proposed evaluation criterion specifically designed for smart contracts has been performed. The proposed evaluation framework has two main categories: Infrastructure criteria that groups all the infrastructure and application related criteria and Development related criteria under Development criteria. The top three smart contract platforms, Ethereum, EOS and Steller have been selected and tested by creating smart contracts on their test environments and subjected them to tests under the proposed evaluation framework. It has been found that Ethereum blockchain provides the best development tools and community support. Stellar provides the easiest UI and smart contract legal considerations. Finally, the EOS blockchain provides the fastest smart contracts and the cheapest costs. On the other hand, it has been found that the Ethereum blockchain has the most expensive costs, and the slowest execution times, whereas Stellar does not have the smart contract flexibility the other blockchains enjoy, and the EOS blockchain has the most convoluted environment and documentation that requires more updates. These findings were in line with the findings of other researchers included in the non-formal research previous work. For future enhancements to the evaluation framework, more criteria could be added such as scalability to improve the comprehensibility of the evaluation framework. Future work on the applicability of the evaluation framework could also include comparison of a select set of specific types of smart contracts or smart contracts on the private or other public blockchains.

Overall, these smart contracts do provide benefits to the layman such as cost and performance, however, stake holders, such as users, investors, and developers have to be aware that smart contracts are hard to develop at the moment. If the aim is to use the smart contracts platform for smart contracts as the definition suggests: "a computerized transaction protocol that executes the terms of a contract" [1], then deploying smart contracts requires technical background with language development skills except if Steller platform was used which does not require expertise.

References

1. Szabo, N.: Smart contracts. Retrieved from http://www.fon.hum.uva.nl/rob/Courses/InformationInSpeech/CDROM/Literature/LOTwinterschool2006/szabo.best.vwh.net/smart.contracts.html (1994)
2. Nakamoto, S.: Bitcoin: A Peer-to-Peer Electronic Cash System (2008)
3. Corbet, S., Lucey, B., Yarovaya, L.: Datestamping the bitcoin and Ethereum bubbles. Financ. Res. Letters **26**, 81–88 (2018)
4. Clack, C.D., Bakshi, V.A., Braine, L.: Smart contract templates: foundations, design landscape and research directions. arXiv preprint arXiv:1608.00771 (2016)
5. Norta, A.: Designing a smart-contract application layer for transacting decentralized autonomous organizations. In: International Conference on Advances in Computing and Data Sciences, pp. 595–604. Springer, Singapore (2016)
6. Wang, S., Yuan, Y., Wang, X., Li, J., Qin, R., Wang, F. Y.: An overview of smart contract: architecture, applications, and future trends. In: 2018 IEEE Intelligent Vehicles Symposium (IV), pp. 108–113. IEEE (2018)
7. Yang, Q., Xu, F., Zhang, Y., Liu, F., Hu, W., Liao, Q.: Design and implementation of a loan system based on smart contract. In: International Conference on Smart Blockchain, pp. 22–31. Springer, Cham (2018)
8. Yang, X., Yi, X., Nepal, S., Han, F.: Decentralized voting: a self-tallying voting system using a smart contract on the Ethereum blockchain. In: International Conference on Web Information Systems Engineering, pp. 18–35. Springer, Cham (2018)
9. Bogner, A., Chanson, M., Meeuw, A.: A decentralized sharing app running a smart contract on the Ethereum blockchain. In: Proceedings of the 6th International Conference on the Internet of Things, pp. 177–178. ACM (2016)
10. Blockgeeks.: A deeper look at different smart contract platforms. Retrieved from https://blockgeeks.com/guides/different-smart-contract-platforms/ (2018)

11. Mulders, M.: Comparison of smart contract platforms. Retrieved from https://hackernoon.com/comparison-of-smart-contract-platforms-2796e34673b7 (2018)
12. Max, T.: Smart contract platforms—Ethereum vs Neo, Lisk, EOS and Cardano. Retrieved from https://www.linkedin.com/pulse/smart-contract-platforms-ethereum-vs-neo-lisk-eos-cardano-mwirabua (2018)
13. Chiu, W.: Smart contract platform comparison (programmer explain). NEO, ETH or ADA. https://medium.com/coinmonks/smart-contract-platform-comparison-programmer-explain-bdc7c303c721 (2018)
14. Macrinici, D., Cartofeanu, C., Gao, S.: Smart contract applications within blockchain technology: a systematic mapping study. Telematics Inform. (2018). https://doi.org/10.1016/j.tele.2018.10.004
15. Delmolino, K., Arnett, M., Kosba, A., Miller, A., Shi, E.: Step by step towards creating a safe smart contract: Lessons and insights from a cryptocurrency lab. In: International Conference on Financial Cryptography and Data Security, pp. 79–94. Springer, Berlin, Heidelberg (2016)
16. Bartoletti, M., Pompianu, L.: An empirical analysis of smart contracts: platforms, applications, and design patterns. In: International Conference on Financial Cryptography and Data Security, pp. 494–509. Springer, Cham (2017)
17. Heitkötter, H., Hanschke, S., Majchrzak, T.A.: Evaluating cross-platform development approaches for mobile applications. In: International Conference on Web Information Systems and Technologies, pp. 120–138. Springer, Berlin, Heidelberg (2012)
18. Cai, W., Wang, Z., Ernst, J.B., Hong, Z., Feng, C., Leung, V.C.: Decentralized applications: the blockchain-empowered software system. IEEE Access **6**, 53019–53033 (2018)
19. Stellar.org.: Retrieved from https://www.stellar.org/ (2019)
20. Atzei, N., Bartoletti, M., Cimoli, T.: A survey of attacks on Ethereum smart contracts (sok). In: Principles of Security and Trust, pp. 164–186. Springer, Berlin, Heidelberg (2017)
21. Golafshani, N.: Understanding reliability and validity in qualitative research. Qual. Rep. **8**(4), 597–606 (2003). Retrieved from https://nsuworks.nova.edu/tqr/vol8/iss4/6
22. Price, C.P., Jhangiani, R., Chiang, I.A.: Research Methods in Psychology, 2nd edn. BCcampus OpenEd, Canada (2015)
23. Michael, R.S.: Measurement: reliability and validity. Y520 Strategies for Educational Inquiry. http://www.indiana.edu/…/week…/reliability_validity_2up (2004)
24. Ropsten.etherescan.: Retrieved from https://ropsten.etherscan.io/ (2019)
25. Stateofthedapps.: Retrieved from https://www.stateofthedapps.com/stats (2019)
26. Chen, T., Li, Z., Zhou, H., Chen, J., Luo, X., Li, X., Zhang, X.: Towards saving money in using smart contracts. In: Proceedings of the 40th International Conference on Software Engineering: New Ideas and Emerging Results, pp. 81–84. ACM (2018)
27. Avan-Nomayo, O.: Ethereum Surpasses Bitcoin in Number of Active Addresses. Retrieved from https://ethereumworldnews.com/ethereum-surpasses-bitcoin-in-number-of-active-addresses/ (2018)
28. Perez, D., Livshits, B.: Smart contract vulnerabilities: does anyone care? arXiv preprint arXiv:1902.06710 (2019)
29. Marino, B., Juels, A.: Setting standards for altering and undoing smart contracts. In: International Symposium on Rules and Rule Markup Languages for the Semantic Web, pp. 151–166. Springer, Cham (2016)
30. Knecht, M., Stiller, B.: SmartDEMAP: a smart contract deployment and management platform. In: IFIP International Conference on Autonomous Infrastructure, Management and Security, pp. 159–164. Springer, Cham (2017)
31. Bragagnolo, S., Rocha, H., Denker, M., Ducasse, S.: SmartInspect: solidity smart contract inspector. In 2018 International Workshop on Blockchain Oriented Software Engineering (IWBOSE), pp. 9–18. IEEE (2018)
32. Mavridou, A., Laszka, A.: Tool demonstration: FSolidM for designing secure Ethereum smart contracts. In: International Conference on Principles of Security and Trust, pp. 270–277. Springer, Cham (2018)

33. Luu, L., Chu, D.H., Olickel, H., Saxena, P., Hobor, A.: Making smart contracts smarter. In: Proceedings of the 2016 ACM SIGSAC Conference on Computer and Communications Security, pp. 254–269. ACM (2016)
34. Porru, S., Pinna, A., Marchesi, M., Tonelli, R.: Blockchain-oriented software engineering: challenges and new directions. In: Proceedings of the 39th International Conference on Software Engineering Companion, pp. 169–171. IEEE Press (2017)
35. Destefanis, G., Marchesi, M., Ortu, M., Tonelli, R., Bracciali, A., Hierons, R.: Smart contracts vulnerabilities: a call for blockchain software engineering? In: 2018 International Workshop on Blockchain Oriented Software Engineering (IWBOSE), pp. 19–25. IEEE (2018)
36. Smart Contracts. Retrieved from https://blockchainhub.net/smart-contracts/ (2019)

Blockchain Languages, Algorithms, Frameworks and Simulation

Blockchain Frameworks

Mohammad Tabrez Quasim, Mohammad Ayoub Khan, Fahad Algarni, Abdullah Alharthy and Goram Mufareh M. Alshmrani

Abstract The blockchain is fascinating in many areas such as finance, healthcare, governance, security and many more. The underlying principle of the blockchain is based on distributed ledgers that solves many existing problems. Every blockchain framework has a different objective and application. There are many important criteria that must be considered during the selection of blockchain framework. Every blockchain framework and development platform has different pros and cons. In this chapter we present some of the leading enterprise frameworks for implementing blockchain solution. The enterprise support, pros and cons and transaction model are few important parameters to select the framework. In this chapter we present detailed investigation on blockchain frameworks that are publicly available.

Keywords Blockchain · PoW · POI · Evaluation framework · Ethereum · Bitcoin · Corda · IOTA · EOS · XRP · Waves · Quorum · NEM · XEM

1 Introduction

The blockchain provides secure computing without involving a central authority in an open network system [1–4]. The companies all around the world are using blockchain technology to enhance the business capacity and Return-on-Investments (ROI). The blockchain has potential to minimize operating cost along with different fraud activities using proof-of-work (PoW). Enterprises are increasingly investing in

M. T. Quasim (✉) · M. A. Khan · F. Algarni · A. Alharthy · G. M. M. Alshmrani
College of Computing and Information Technology, University of Bisha, Bisha, Saudi Arabia
e-mail: mtabrez@ub.edu.sa; ayoub.khan@ieee.org

M. A. Khan
e-mail: ayoub.khan@ieee.org

F. Algarni
e-mail: fahad.alqarni@ub.edu.sa

G. M. M. Alshmrani
e-mail: gmoubark@ub.edu.sa

M. A. Khan et al. (eds.), *Decentralised Internet of Things*, Studies in Big Data 71,
https://doi.org/10.1007/978-3-030-38677-1_4

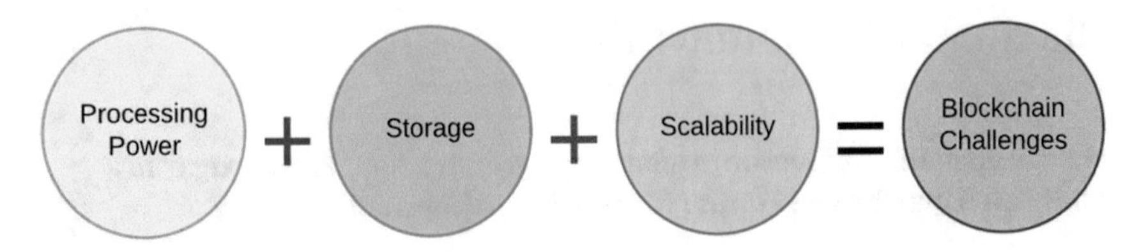

Fig. 1 Blockchain computational complexity

blockchain technology for extremely secure and transparent mechanism for tracking ownership of assets across the boundaries.

Definition The blockchain framework can be defined as software solutions that simplifies the development and deployment of blockchain applications with little customization.

The blockchain frameworks contains infrastructure and libraries to develop the application. The network infrastructure or simply infrastructure consist of nodes and software running on them. The node can be physical machine, virtual machine or containers. The software providing features and capability such as user identity, transaction details, consensus protocol, also controls the identity management for blockchains. The application is composed of the code running inside the infrastructure known as smart contract in general. The client application interacts with the infrastructure. The latter serves as the interface or access point from external world. A good blockchain framework shall allow application development outside the actual deployment of the network. We don't need to deploy the network before developing the application because, if our application can run on a small network, it should run as well in a bigger infrastructure or actual deployment environment.

The selection of an enterprise blockchain framework is a little tricky affair, since any particular framework don't provide all the features. The main challenges in any blockchain framework are processing power, storage and scalability as shown in Fig. 1. The stakeholders must be very careful about variety of factors before, during, and after implementation of blockchain in the enterprise.

There are many important criteria that must be considered during the selection of blockchain framework. We present some of the criteria for selection of blockchain framework as shown in Table 1.

Different blockchain frameworks and development platforms has different pros and cons. In next section we will present some of the leading enterprise frameworks for implementing blockchain solutions.

2 Ethereum

Ethereum is open source framework where we can create virtually any decentralised online services on the DApps that operate on the basis of smart contracts. The concept

Table 1 Selection criteria for blockchain framework

Criteria	Description
License	Type of licensing viz free or paid and features
Support model	Framework support, longevity, popularity
Activity	Upgrade support for framework
Roadmap	Vision and roadmap of the framework
Ease of use	large-scale adoption and intuitive
Reliable backing	Open source community or corporate community

of Ethereum was proposed by Vitalik Buterin in 2013, but it was only possible to implement it in 2015 [5–7]. The Ethereum [1, 8, 9] is based on four important key components as below:

2.1 Ethereum Virtual Machine (EVM)

The EVM is a quasi-Turing machine that has EMV bytecode [1]. The EVM is quasi turning machine because computations performed by the machine are bound by gas, that limits the number of computations. This is the also called "full Turing environment" in which we can execute DApps written in one of several popular programming languages. In other words, instead of creating a separate framework for each language or application, they all work on the same blockchain. This makes the Dapps development process efficient and simple.

2.2 Smart Contracts

Smart contract is executed automatically subject to certain conditions specified in the executable code as shown in Fig. 2. Therefore, we can say that smart contract is an executable program [4]. The smart contract is a computer algorithm for the exchange of cryptocurrency, real estate, gold or other value without the participation of third parties or guarantors.

2.3 Decentralised Applications (DApps)

Decentralised applications that use smart contracts for various purposes: putting digital signatures, forecasting stock markets, guaranteeing the transfer of valuables, and the like. More than half of live DApps –Ethereum applications [10].

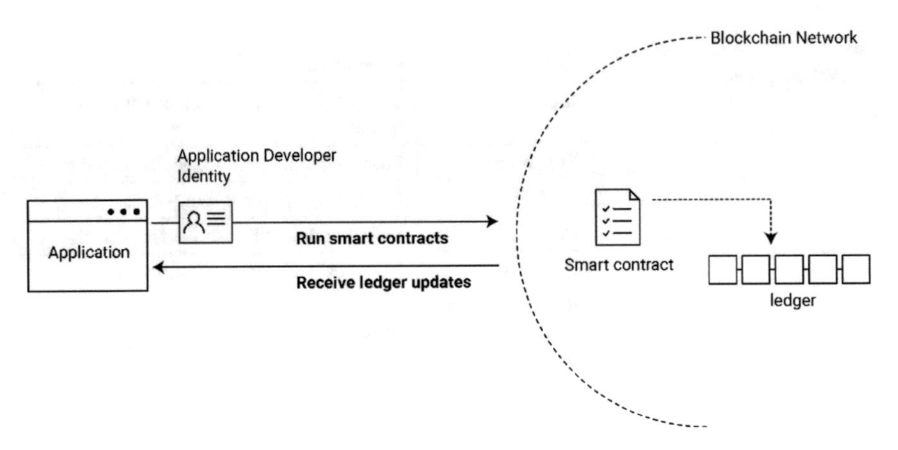

Fig. 2 Smart contract execution

2.4 *Performance*

There are many parameters to evaluate the performance of the framework. The Ethereum uses Merkle trees to optimize transactional hashing and increase potential for scalability. The Ethereum is the most popular framework as 80% of the project are being developed in Ethereum [11–13]. The success of Ethereum is due to the fact that this is the first platform on which full-fledged smart contracts are implemented. Also, Ethereum is relatively quick and easy to launch Initial Coin Offering (ICO). We summarise some of the important pros and cons of Ethereum in Table 2.

3 Hyperledger

The Hyperledger is supported by Linux Foundation and IBM that is an open source collaborative effort which is particularly useful for advancing cross-industry blockchain technologies [14–16]. Hyperledger is a global initiative that includes

Table 2 Pros and cons of ethereum

Pros	Cons
Leverages blockchain	Sluggish transaction speed
High capitalization: $9.7 billion; Supported by big players like IBM, Microsoft, JPorgan Chase, Amazon and others;	Centralization—hacking of the DAO showed that the word of developers is more important than community voting
Strong team	First-mover disadvantage
Proven reliability	Market hoaxes
Per day traffic	PoS update

Table 3 Components of hyperledger

Component	Description
Ledger	An append-only distributed ledger
Consensus algorithm	A consensus algorithm for agreeing to changes in the ledger
Privacy	Privacy of transactions through permissioned access
Smart contracts	Smart contracts to process transaction requests

industry leaders from banking, finance, manufacturing, supply chain, Internet of Things, and other technology. The Hyperledger project has started in 2015 and more than 100 companies participate in Hyperledger [15, 16]. The Hyperledger has very strong industry backup by many financial institutions such as JP Morgan, London Stock Exchange (LSE), Deutsche Boerse and CME, SWIFT, Moscow Exchange, ABN AMRO, BNP Paribas, Wells Fargo and hundreds of other companies also joined the Linux Foundation project. From IT giants, the Hyperledger has technical backup by Cisco, IBM, Microsoft, Fujitsu and Intel [17–19].

3.1 *Components of Hyperledger Frameworks*

Hyperledger business blockchain frameworks are used to build enterprise blockchains. The components shown in Table 3 are essential to build Hyperledger framework.

3.2 *Example of Hyperledge Framework*

There are many frameworks available for Hyperledger as summarised in Table 4.

Table 4 Examples of hyperledger framework

Framework	Contributor	Domain	Consensus algorithm	Development environment
Fabric	IBM	Smart contracts	BFT	Java, JavaScript
Sawtooth	Intel	tokenize logistics and sales chains	PoET	Any language
Burrow	Monax	Smart contract	PoET	Solidity
Iroha	Soramitsu etc.	Mobile applications	YAC	C++
Indy	Sovrin	Digital identities	RBFT	Python, Java, C#, node.js

Table 5 Pros and cons of hyperledge framework

Pros	Cons
Modular architecture	Complex fabric architecture
Hybrid model	Network fault tolerance
Optimized performance and scalability	Lack of proven use cases
SQL-like query capability	Inadequately skilled programmers
Permission membership:	Network fault tolerance

Below are few pros and cons of the Hyperledger framework as shown in Table 5.

4 Bitcoin

The Bitcoin is the first and most prominent cryptocurrency in the financial world. A group of people under the common alias Satoshi Nakamoto form the Bitcoin in year 2009 [20–24]. The Bitcoin blockchain is the progenitor of a significant part of the first twenty of cryptocurrencies: Ethereum, Litecoin, Dash, Bitcoin Cash, Bitcoin SV and others [25–27]. This is also very interesting to know that to date nobody knows who is the owner of the company, for example if we do search for domain details on who is lookup then we will get below information as shown in Fig. 3.

The size of blockchain is also increasing day by day as shown in Fig. 4.

The Bitcoin has the largest capitalization and subscription. Some important pros and cons of Bitcoin have been listed in Table 6.

5 Corda

Corda is a blockchain platform primarily developed for legal contracts and other shared data between mutually trusting organizations, however, this makes it possible for a diverse range of applications to interoperate on a single network [28, 29]. The Corda is platform operate within the Java Virtual Machine (JVM). Corda was established by the R3 consortium (R3CEV LLC) for recording, monitoring and synchronizing financial agreements between regulated financial institutions. The consensus algorithm uses "notarized" nodes to verify and sign contracts. Transaction information is not broadcast to all network nodes i.e. restricted. The information is made available only to nodes that have confirmed legitimate interests in those assets that participate in the transaction. If the transaction is between A and B, then only the nodes of these two structures will receive information.

The Corda's has great potential to scale as compared to other frameworks. We present some of the important pros and cons of Cordra in Table 7 [29].

Domain Name: BITCOIN.ORG
Registry Domain ID: D153621148-LROR
Registrar WHOIS Server: whois.namecheap.com
Registrar URL: http://www.namecheap.com
Updated Date: 2018-09-25T15:44:30Z
Creation Date: 2008-08-18T13:19:55Z
Registry Expiry Date: 2021-08-18T13:19:55Z
Registrar Registration Expiration Date:
Registrar: NameCheap, Inc.
Registrar IANA ID: 1068

Registrar Abuse Contact Email: abuse@namecheap.com
Registrar Abuse Contact Phone: +1.6613102107
Reseller:
Do-
main Status: clientTransferProhibited https://icann.org/epp#clientTransf
erProhibited
Registrant Organization: WhoisGuard, Inc.
Registrant State/Province: Panama
Registrant Country: PA
Name Server: DNS1.REGISTRAR-SERVERS.COM
Name Server: DNS2.REGISTRAR-SERVERS.COM
DNSSEC: unsigned
URL of the ICANN Whois Inaccuracy Complaint Form https://www.ican
n.org/wicf/)

For more information on Whois status codes, please visit https://icann.o
rg/epp

Fig. 3 WHOIS domain information

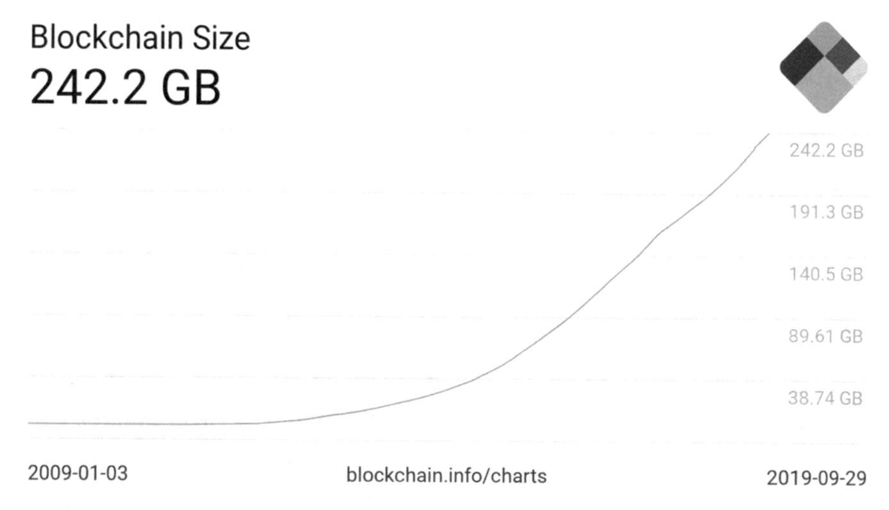

Fig. 4 Blockchain size of bitcoin [4]

Table 6 Pros and cons of BITCOIN

Pros	Cons
Payment freedom	Risk and volatility
Highly capitalised	Slow speed (7 transaction/sec)
Control and security	Blockchain size—242.2 GB
High volume trading	Lack of awareness & understanding
Low commission fee	No full-fledged smart contracts

Table 7 Pros and cons of corda

Pros	Cons
The ability to search for consensus for individual contracts and agreements;	The use of oracles (people) to confirm the authenticity of documents and information reduces the legitimacy of these checks
Restricted access	Corda framework customized to the financial sector
Built on industry-standard tool R3	Limited user only in financial sector

6 EOS

EOS blockchain platform is built for both public and private use cases. The EOS is well suited for business needs across industries which requires role-based security permissions, industry-leading speeds and secure application processing [30]. Almost all the decentralised applications can be launched using EOS. The EOS managed to collect 465 million dollars during the ICO [30].

The investor has the right in the income, property, copyrights, reputation in the proportion of investment or token acquired. The holder of tokens may get access to DApps developed under the new project, storage access, received divided etc. like a stake holder in any company. We present some important pros and cons of EOS in Table 8.

Table 8 Pros and cons of EOS

Pros	Cons
Throughput—1200 operations per second	Project in development-there is no even a GUI wallet
High potentials for scaling due to delegated proof-of-stake consensus algorithm	Crowdfunding conditions may frighten DApps developers;
Innovative crowdfunding model	EOS team is in no hurry to report to investors
Convenient DApps development toolkit	The most criticized cryptocurrency of the top ten

7 IOTA

This is new transaction settlement for the Internet of Things applications. It introduced a new way to perform transaction through a peer to peer system, called a tangle. The IOTA framework was released in 2016 and has a market capital more than $10 billion [31–33]. The IoT devices perform PoW. The IOTA doesn't support smart contracts [34, 35].

Unlike the other frameworks such as Ethereum, Bitcoin, the IoTA has no traditional blockchain structure [36]. The IOTA uses the Tangle algorithm in which transaction will be confirm from two other users.

The tangle is based on Directed Acyclic Graph (DAG) as shown in Fig. 5. The tangle data structure moves in one direction without looping back onto itself.

The time elapse from left to right in this graph. Each circle represents a transaction issued by a device on the network. In blockchain, there are two distinct types of participants in the system, those who issue transactions, and those who approve transactions, however, in the tangle, every device or node works to maintain the ledger. Every node is also a kind of miner. When a node wants to transfer some value, then it must validate two previous transactions, which is shown by the arrows in the Fig. 5. The validation of transaction requires a small amount PoW in order to secure the network. Since, the node are miners as well therefore no transaction fees is applicable. We discuss some of important pros and cons of IoTA are discuss in Table 9.

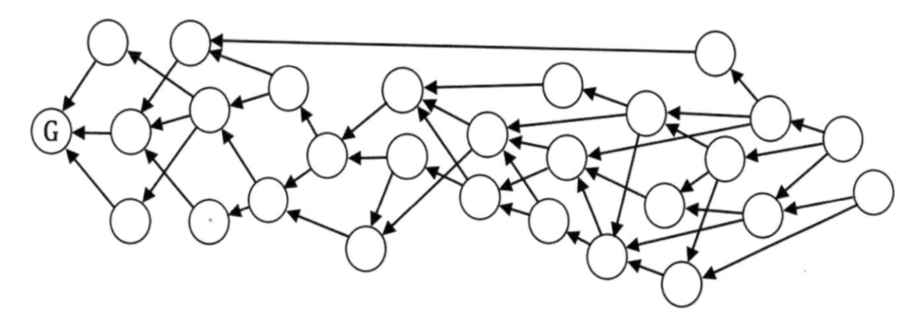

Fig. 5 Tangle as direct acrylic graph

Table 9 Pros and cons of IoTA

Pros	Cons
Micro-payments	Ternary based logic
Scalability	There are hidden areas in the code
Lightweight	Framework functions are limited to IoT
Quantum-secure	No support of smart contracts

Table 10 Pros and cons of XRP

Pros	Cons
Capacity—1000 operations per second;	Ripple Labs holds 65% of platform tokens;
Transaction fee—from 0.00001 XRP	Practice has shown that the project team can block user tokens if they don't like it
Support for large banks	Complex architecture
Ability to cancel transactions	Less scalable

8 Ripple (XRP)

The XRP is formerly known as OpenCoin Ripple [35]. The OpenCoin is built for payment and exchange network known as RippleNet which is on the top of a distributed ledger database called as XRP Ledger. The objective of XRP is to connect banks, payment providers and digital asset exchanges, enabling faster and cost-efficient global payments. The XRP was first idealized in 2004 by Ryan Fugger, who developed the first prototype of Ripple known as RipplePay in 2005 [35]. In 2012, Jed McCaleb and Chris Larsen et al. founded the US-based technology company OpenCoin. The XRP Ledger is not based on a PoW consensus algorithm, instead, it uses a customised algorithm known as the Ripple Protocol Consensus Algorithm (RPCA). The XRP was ranked second in terms of capitalization, displacing Ethereum from $ 9.5 billion to $12.3 billion as of December 10, 2018. This happened against the background of another sinking of the market, which indicates a great confidence of investors and users to the Ripple blockchain. In Table 10, we present some important pros and cons of Ripple framework [35].

9 Waves

Waves blockchain framework is an open, decentralised blockchain platform where anyone can build application along with new cryptocurrency. The Waves Platform was founded by Sasha Ivanov, a Russian physicist. The unique feature of Waves is to create an infrastructure that makes it easy for application developers to create all kinds of blockchain-based applications. The Waves is a software framework consists of different tools and utilities to help developers build applications. In order to run those applications, users pay fees using the WAVES tokens. Unlike Bitcoin, Waves platform has decentralised exchange (DEX). The DEX allows users to trade bitcoin, WAVES or any token issued on the Waves platform directly on a peer-to-peer level. The DEX has fast transactions, low fees and secure settlements on the blockchain. If we want to issue our own crypto coin then you need minimum one Wave token. Anyone can buy WAVES token with bitcoin on several cryptocurrency exchanges such as Bittrex, Tidex, and YoBit. One of the important features which distinguish WAVES with other

Table 11 Pros and cons of WAVES

Pros	Cons
Ideal for crowdfunding	Incomprehensible legal status of project on waves
Fiat transfers	Support issue
Highly accessible	Exchange is yet not mature
Fast, low-cost and scalable	Performance issue

frameworks is direct exchange between fiat currencies, cryptocurrencies and real-world commodities through the decentralised exchange. The WAVES framework is implemented using Python, that makes it easy for developers to implement and integrate new applications using APIs through JSON-RPC, REST etc. We have listed some of the Pros and Cons of Waves are discuss in Table 11.

10 Quorum

The Quorum was developed by JP Morgan to address key issues in the financial industry through a distribution registry and smart contract (Ethereum) capabilities. The Quorum has generated institutional transaction volumes. The Quorum is able to restrict access to transaction history with transparency of the system.

Quorum uses Raft/Istanbul BFT consensus algorithms. When a new transaction appears then it is routed to main node that directs further to other nodes and requesting confirmation of authenticity without any communication. Quorum has focus on enterprise, permissioned blockchain to cover all the use cases of financial sector. Quorum is built on Ethereum. The components of Quorum are shown in Fig. 6.

Like every framework the Quorum has also some pros and cons as shown in Table 12.

11 NEM (XEM)

The (New Economy Movement) NEM platform is designed from scratch to gain high scalability and speed [35]. The NEM is a permissioned private blockchain with a revolutionary consensus mechanism called Proof-of-Concept (POI). The POI is the mechanism that is used to determine which network participants are more important, thereby, eligible to add a block to the blockchain. The NEM uses 'harvesting' process to determine the important participant. The accounts with a higher importance score will have a higher probability of being chosen to harvest a block. To become eligible the participant must hold at least 10,000 vested XEM.

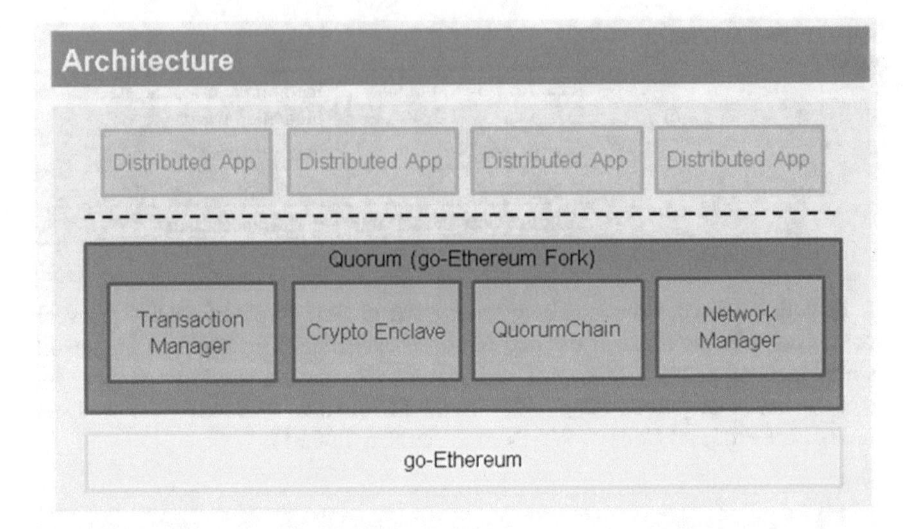

Fig. 6 Quorum architecture

Table 12 Pros and cons of quorum

Pros	Cons
High performance	Limited use of capabilities of the framework
Enhanced transaction and contract privacy	Transactions are confirmed by several "selected" nodes
Voting-based consensus mechanisms	Messaging overhead
Better performance	Limited block size
permissioned blockchain	Anonymity

The architectural design of NEM is open and scalable. The NEM Platform has focus on trading, banking and charity. The NEM uses a different consensus algorithm called Proof-of-Importance (PoI) [35].

The low transaction confirmation fee, energy consumption and performance are major advantages of NEM. The NEM provides RESTful JSON API interface for application development [35]. We present some of the pros and cons of NEM in Table 13.

12 Conclusion

The blockchain technology has gained considerable popularity in the last couple of years. The blockchain framework provides an infrastructure and environment

Table 13 Pros and cons of NEM [35]

Pros	Cons
Good capacity	Centralisation POI
Low commission	Weak debut
Less energy consumption	Low NEM consumer
Anti-counterfeiting	Less literature

to develop any blockchain application. We presented many popular frameworks of blockchain technology; some frameworks solve problems related to specific areas such as IoT, Smart contract, cryptocurrency and so on. We discussed basic features, components, characteristics, pros and cons of Ethereum, Hyperledger, Bitcoin, Corda, EOSIOTA, XRP, Waves, Quorum, and NEM. The enterprise support, pros and cons and transaction are few important parameters to select the framework. However, this is a challenging task to select the blockchain framework due to volatility in the customer choice and market dynamics.

References

1. Raikwar, M., Mazumdar, S., Ruj, S., Sen Gupta, S., Chattopadhyay, A., Lam, K.: A blockchain framework for insurance processes. In: 2018 9th IFIP International Conference on New Technologies, Mobility and Security (NTMS), pp. 1–4. Paris (2018). https://doi.org/10.1109/ntms.2018.8328731
2. Puthal, D., Malik, N., Mohanty, S.P., Kougianos, E., Yang, C.: The blockchain as a decentralized security framework [Future Directions]. IEEE Consum. Electron. Mag. **7**(2), 18–21 (2018). https://doi.org/10.1109/mce.2017.2776459
3. Xu, Y., Wu, M., Lv, Y., Zhai, S.: Research on application of block chain in distributed energy transaction. In: 7th IEEE 3rd Information Technology and Mechatronics Engineering Conference (ITOEC), pp. 957–960. Chongqing (2017). https://doi.org/10.1109/itoec.2017.8122495
4. Blockchain. https://www.blockchain.com/en/charts/blocks-size?timespan=all. Accessed 29 Sep 2019
5. Pinna, A., Ibba, S., Baralla, G., Tonelli, R., Marchesi, M.: A massive analysis of ethereum smart contracts empirical study and code metrics. IEEE Access **7**, 78194–78213 (2019). https://doi.org/10.1109/access.2019.2921936
6. Luo, X., Cai, W., Wang, Z., Li, X., Victor Leung, C.M.: A payment channel based hybrid decentralized ethereum token exchange. In: 2019 IEEE International Conference on Blockchain and Cryptocurrency (ICBC), pp. 48–49. Seoul, Korea (2019). https://doi.org/10.1109/bloc.2019.8751454
7. di Angelo, M., Salzer, G.: A survey of tools for analysing ethereum smart contracts. In: 2019 IEEE International Conference on Decentralized Applications and Infrastructures (DAPPCON), pp. 69–78. Newark, CA, USA (2019). https://doi.org/10.1109/dappcon.2019.00018
8. Augusto, L., Costa, R., Ferreira, J., Jardim-Gonçalves, R.: An application of ethereum smart contracts and IoT to logistics. In: 2019 International Young Engineers Forum (YEF-ECE), pp. 1–7. Costa da Caparica, Portugal (2019). https://doi.org/10.1109/yef-ece.2019.8740823

9. Aung, Y.N., Tantidham, T.: Review of ethereum: smart home case study. In: 2017 2nd International Conference on Information Technology (INCIT), pp. 1–4. Nakhonpathom (2017). https://doi.org/10.1109/incit.2017.8257877
10. Pierro, G.A., Rocha, H.: The influence factors on ethereum transaction fees. In: 2019 IEEE/ACM 2nd International Workshop on Emerging Trends in Software Engineering for Blockchain (WETSEB), pp. 24–31. Montreal, QC, Canada (2019). https://doi.org/10.1109/wetseb.2019.00010
11. Tantidham, T., Aung, Y.N.: Emergency service for smart home System using ethereum blockchain: system and architecture. In: 2019 IEEE International Conference on Pervasive Computing and Communications Workshops (PerCom Workshops), pp. 888–893. Kyoto, Japan (2019). https://doi.org/10.1109/percomw.2019.8730816
12. Harris, C.G.: The risks and challenges of implementing ethereum smart contracts. In: 2019 IEEE International Conference on Blockchain and Cryptocurrency (ICBC), pp. 104–107. Seoul, Korea (South) (2019). https://doi.org/10.1109/bloc.2019.8751493
13. Dika, A., Nowostawski, M.: Security vulnerabilities in ethereum smart contracts. In: 2018 IEEE International Conference on Internet of Things (iThings) and IEEE Green Computing and Communications (GreenCom) and IEEE Cyber, Physical and Social Computing (CPSCom) and IEEE Smart Data (SmartData), pp. 955–962. Halifax, NS, Canada (2018). https://doi.org/10.1109/cybermatics_2018.2018.00182
14. Attia, O., Khoufi, I., Laouiti, A., Adjih, C.: An Iot-blockchain architecture based on hyperledger framework for healthcare monitoring application. In: 2019 10th IFIP International Conference on New Technologies, Mobility and Security (NTMS), pp. 1–5. CANARY ISLANDS, Spain (2019). https://doi.org/10.1109/ntms.2019.8763849
15. Goranović, A., Meisel, M., Wilker, S., Sauter, T.: Hyperledger fabric smart grid communication testbed on raspberry PI ARM architecture. In: 2019 15th IEEE International Workshop on Factory Communication Systems (WFCS), pp. 1–4. Sundsvall, Sweden (2019). https://doi.org/10.1109/wfcs.2019.8758000
16. Nguyen, T.S.L., Jourjon, G., Potop-Butucaru, M., Thai, K.L.: Impact of network delays on hyperledger fabric. In: IEEE INFOCOM 2019—IEEE Conference on Computer Communications Workshops (INFOCOM WKSHPS), pp. 222–227. Paris, France, (2019)
17. Ampel, B., Patton, M., Chen, H.: Performance modeling of hyperledger sawtooth blockchain. In: 2019 IEEE International Conference on Intelligence and Security Informatics (ISI), pp. 59–61. Shenzhen, China (2019). https://doi.org/10.1109/isi.2019.8823238
18. Thakkar, P., Nathan, S., Viswanathan, B. Performance benchmarking and optimizing hyperledger fabric blockchain platform. In: IEEE 26th International Symposium on Modeling, Analysis, and Simulation of Computer and Telecommunication Systems (MASCOTS), pp. 264–276. Milwaukee, WI (2018). https://doi.org/10.1109/mascots.2018.00034
19. Ampel, B., Patton, M., Chen, H.: Performance modeling of hyperledger sawtooth blockchain. In: 2019 IEEE International Conference on Intelligence and Security Informatics (ISI), pp. 59–61. Shenzhen, China (2019). https://doi.org/10.1109/isi.2019.8823238
20. Lin, Y., Wu, P., Hsu, C., Tu, I., Liao, S.: An evaluation of bitcoin address classification based on transaction history summarization. In: 2019 IEEE International Conference on Blockchain and Cryptocurrency (ICBC), pp. 302–310. Seoul, Korea (South) (2019). https://doi.org/10.1109/bloc.2019.8751410
21. Qin, K., Hadass, H., Gervais, A., Reardon, J.: Applying private information retrieval to lightweight bitcoin clients. In: Crypto Valley Conference on Blockchain Technology (CVCBT), pp. 60–72. Rotkreuz, Switzerland (2019). https://doi.org/10.1109/cvcbt.2019.00012
22. Grundmann, M., Leinweber, M., Hartenstein, H.: Banklaves: concept for a trustworthy decentralized payment service for bitcoin. In: 2019 IEEE International Conference on Blockchain and Cryptocurrency (ICBC), pp. 268–276. Seoul, Korea (South (2019). https://doi.org/10.1109/bloc.2019.8751394
23. Mittal, A., Dhiman, V., Singh, A., Prakash, C.: Short-term bitcoin price fluctuation prediction using social media and web search data. In: 2019 Twelfth International Conference on Contemporary Computing (IC3), pp. 1–6. Noida, India (2019). https://doi.org/10.1109/ic3.2019.8844899

24. Biryukov, A., Tikhomirov, S.: Transaction clustering using network traffic analysis for bitcoin and derived blockchains. In: IEEE INFOCOM 2019—IEEE Conference on Computer Communications Workshops (INFOCOM WKSHPS), pp. 204–209. Paris, France (2019). https://doi.org/10.1109/infcomw.2019.8845213
25. Chadha, G.K., Singh, A.: Bitcoin block-chain mining. In: 2019 9th International Conference on Cloud Computing, Data Science & Engineering (Confluence), pp. 152–157. Noida, India (2019). https://doi.org/10.1109/confluence.2019.8776961
26. Erfani, S., Ahmadi, M.: Bitcoin security reference model: an implementation platform. In: 2019 International Symposium on Signals, Circuits and Systems (ISSCS), pp. 1–5. Iasi, Romania (2019). https://doi.org/10.1109/isscs.2019.8801796
27. Wu, Y., Luo, A., Xu, D.: Forensic analysis of bitcoin transactions. In: 2019 IEEE International Conference on Intelligence and Security Informatics (ISI), pp. 167–169. Shenzhen China (2019). https://doi.org/10.1109/isi.2019.8823498
28. R3 Corda. https://docs.corda.net/releases/release-M0.1/data-model.html. Accessed 11 Sep 2019
29. Richard Gendal Brown. The corda platform: an introduction (2018). https://www.corda.net/content/corda-platform-whitepaper.pdf. Accessed 1 Sep 2019
30. Sing, N.: Top 7 benefits of EOS blockchain (2018). https://101blockchains.com/top-7-benefits-of-eos-blockchain/. Accessed 1 Sep 2019
31. Lamtzidis, O., Gialelis, J.: An IOTA based distributed sensor node system. In: 2018 IEEE Globecom Workshops (GC Wkshps), pp. 1–6. Abu Dhabi, United Arab Emirates (2018). https://doi.org/10.1109/glocomw.2018.8644153
32. Dasalukunte, D., Mehmood, S., Öwall, V.: Complexity analysis of IOTA filter architectures in faster-than-Nyquist multicarrier systems. In: 2011 NORCHIP, pp. 1–4. Lund (2011). https://doi.org/10.1109/norchp.2011.6126704
33. Shabandri, B., Maheshwari, P.: Enhancing IoT security and privacy using distributed ledgers with IOTA and the tangle. In: 2019 6th International Conference on Signal Processing and Integrated Networks (SPIN), pp. 1069–1075. Noida, India (2019). https://doi.org/10.1109/spin.2019.8711591
34. Glencross, M., Howard, T., Pettifer, S.: Iota: an approach to physically-based modelling in virtual environments. In: Proceedings IEEE Virtual Reality 2001, pp. 287–288. Yokohama, Japan (2001). https://doi.org/10.1109/vr.2001.913800
35. Tarasenko, E.: Best blockchain frameworks you should know about. https://merehead.com/blog/blockchain-frameworks-you-should-know-about/ Accessed 2 July 2019
36. Andriopoulou, F., Orphanoudakis, T., Dagiuklas, T.: IoTA: IoT automated SIP-based emergency call triggering system for general eHealth purposes. In: 2017 IEEE 13th International Conference on Wireless and Mobile Computing, Networking and Communications (WiMob), pp. 362–369. Rome (2017). https://doi.org/10.1109/wimob.2017.8115830

Consensus Algorithm

Rashmi Bhardwaj and Debabrata Datta

Abstract Consensus algorithm in general is framed as a decision-making process where a group of people express their individual opinions to construct the decision which provides a best estimate of a process or system. Each member of the group expresses their opinion to support the decisions taken for a course of action. In simple terms, it is just a method to decide any event to occur within a group. Every one present in the group can suggest an idea, but the majority will be in favor of the one that helps them the most. Others have to deal with this decision whether they liked it or not. Byzantine Fault Tolerance (BFT), a problem of Byzantine General, is a system with a particular event of failure. One can experience best the aforementioned situation (BFT) with a distributed computer system. Many times, there can be malfunctioning consensus systems. These components are responsible for the further conflicting information. Consensus systems can only work successfully if all the elements work in harmony. However, if even one of the components in this system malfunctions the whole system could break down. These Blockchain consensus models are just the way to reach an agreement. However, there can't be any decentralized system without common consensus algorithms. It won't even matter whether the nodes trust each other or not. They will have to go by certain principles and reach a collective agreement. In order to do that, it is required to check out all the Consensus algorithms. It can be stated that versatility of blockchain networks is due to consensus algorithms. However, blockchain consensus algorithm may have pros and cons which can always alter the perfection of the algorithm.

Keywords Consensus Algorithm (CA) · Artificial Intelligence (AI) · Cognitive Intelligence (CI) · Human Intelligence (HI) · Blockchain · Peer-to-peer network (P2P)

R. Bhardwaj (✉)
Nonlinear Dynamics Research Lab, University School of Basic & Applied Sciences,
Guru Gobind Singh Indraprastha University, Delhi, India
e-mail: rashmib@ipu.ac.in

D. Datta
Radiological Physics & Advisory Division, Bhabha Atomic Research Centre,
CT & CRS Building, Anushakti Nagar, Mumbai, India
e-mail: ddatta@barc.gov.in

M. A. Khan et al. (eds.), *Decentralised Internet of Things*, Studies in Big Data 71,
https://doi.org/10.1007/978-3-030-38677-1_5

1 Introduction

With a view to recent advancement of Artificial Intelligence (AI), Cognitive intelligence (CI), Human Intelligence (HI) and data science, consensus plays a major role in formulating information in an ordered manner so that a substantial amount learning in the form of a packet of knowledge can be gained. Basically, a problem that arises in distributed systems and replicates a common state, for example, the data/schema depicted in a database is called as consensus. In view of this fact,one can define consensus algorithm as a methodology or schema in the field of computer science that can be used to derive a single or unique agreeable opinion with respect to a data point (singleton) within a processes which are widely distributed. In order to quantify the safety or validity of a data network consisting of multiple unreliable nodes, consensus algorithms (CA) play an essential role. Decision making for an event to qualify after passing through rigorous tests on the basis of opinions from a group of experts, consensus can be categorized as evidences aggregated using certain procedures. Conflict resolution that takes place among various experts is resolved using Dempster-Shafer algorithm. Distributed computing and multi-agent systems always invite consensus algorithms. Consensus algorithm works under the hypothesis of non-availability of certain processes (may be systems if algorithm deals with engineering structural systems). Under the hypothesis of failure of certain communications algorithm works. This characteristic feature of consensus algorithm is generally observed for accommodating the issues for any multi-agent systems. In this context, we can say that consensus algorithms are fault-tolerant. Consensus algorithms can be applied to carry out tasks which are (1) Decision of making a commitment of a distributed data processing to a database, (2) Designation of nodal points for acceptance of a distributed data processing as leader and (3) Synchronizing replicates of state machine ensuring consistency within processors. As an example, consensus algorithm supports many systems available in the world such as load balancing systems, systems having smart grids, systems for synchronization of clock and control of drone. In multiple unreliable asynchronously connected replicas, consensus algorithms deal for keeping the state consistent.

Literal meaning of consensus is "an accepted opinion or agreement on a single item networked from many resources as decisions". A large number of domain experts engaged in making these decisions is known as decision makers. Decision making can be of two groups. One group in which decisions are based on more than one criterion is known as multi-criteria decision making (MCDM) process. In this group many experts express their individual opinion on all the criteria and finally all those opinions for particular criteria are aggregated by using aggregation operator. Other group of decision making methodology is based on multi attribute known as multi attribute decision making (MADM). For example, to select a site for launching a chemical industry in a public domain several people's decisions are taken. A consensus algorithm can be based on either MCDM or MADM generally used to achieve a particular event resourced from several domain experts. In the field of information technology, we can classify this algorithm as a distributed process. If the decisions are originated in terms of information networked in a distributed computing platform, then authenticity is one of the main issues to accept each and every decision at a common point. Various resources from where decisions are originated may be labeled as unreliable nodes. If we say that the agreement on the net outcome of the event is reliable, then the reliability of the agreeable outcome is achieved by consensus algorithms (MCDM and MADM). In short, consensus is a methodology of an agreement of an outcome among a group of outcome originated from many resources. As resources of information in MCDM or MADM are sparse, processing data in this case is known as distributed data processing. However, in the field of distributed data processing very often, failure occurs because at proper time required data may not reach at identified location. Alternatively, we can say that data processing in a distributed network becomes extremely difficult due to failure of the data communication from various resources (nodes or objects encapsulated with all varieties). Literature survey [1–4] notifies that applications of consensus algorithms are: (a) decision making of commitment of a distributed transaction in a database, (b) identifying a leadership node for distributed task, and (c) synchronization of state machine replicas.

Consensus procedures also allow safety in updating a distributed shared state and provide very good research topics in the field of artificial intelligence and data science. A fault tolerant (FT) is defined to be a system which provides its service in presence of faults [4]. In order to make a distributed system as FT, the states are distributed and shared across many such similar states across the network. These shared states are updated by predefined state transition rules and they are executed subsequently on all the replicas. This technique is said to be replication of state machine [1]. It can be also noted from that there is no loss of the provided one or more nodes crash ensured by replication of state. It is also known that there is a guarantee about operation of each piece of information or data (nodes) with similar inputs by rules of the state machine. Achievement of consensus for a distributed system is interesting as well as challenging. Various type of tasks associated with a distributed system such as node failure, network partitioning, time delay of messages, out of order of messages during their launching, and also messages corrupted, should be resilient by consensus algorithms. Consensus algorithms also have to tackle selfish and improvised malicious piece of information. In order to solve the tasks associated with a distributed system, a large number of algorithms are published in the literature [1]. Every algorithm makes the appropriate assumptions in terms of synchrony [1]. A short review of available various consensus algorithms will allow users to understand its utility in the field of data science. In view of the literature study, we can say that consensus algorithms support a large number of practical applications such as balancing of load in an engineering structure, smart grid distribution of pay load, unmanned vehicle like control of drone, etc. Among all consensus algorithm, the first consensus algorithm "Paxos" can be found elsewhere in [2]. As per literatures [2, 3] we found that agreement of an event from various decisions (nodes) by consensus algorithm generates the decentralized database technology. It has been observed the piece information, message received or data sent from distributed nodes to a designated point are precise; however in practice these items are generally imprecise or incomplete, in a common sense, data are corrupted. Therefore, with corrupt data processing in a distributed network becomes much more complicated. In order to handle this situation we have developed a new consensus algorithm EVICON on the basis of soft computing technique. In general soft computing is based on fuzzy or corrupted data, however EVICON is mainly based on possibilistic handling of fuzzy data. This chapter presents a short review of available consensus algorithms in Sect. 2. Section 3 describes EVICON in detail with various case studies. Finally, Sect. 4 will draw the conclusions for this chapter.

2 Review of Consensus Algorithms

This section provides a short review of available consensus algorithms including their pros and cons.

2.1 *Blockchain Consensus Algorithm*

A blockchain is a growing list of records. Each piece of records is called as blocks. All these blocks in a blockchain are connected by cryptography [4]. A blockchain can be also considered as a database (blocks) which is sparse and managed by a large number of computers connected on a network known as peer-to-peer (P2P) network [4, 5]. To make a distributed ledger, blockchain plays a major role. A blockchain is typically managed by a P2P network. A blockchain is tuned collectively to a protocol for inter-node communication [5]. A single point of failure is prevented by a copy of the ledger maintained in P2P [5]. Updation with validations is reflected in all copies simultaneously. Security and safety in an unreliable network is achieved by Bitcoin which uses the proof of work algorithm [5]. Encapsulation is taken into account to design a blockchain and due to this design modification of the data can be ruled out. It is a distributed ledger and open. Transactions of records between two parties take place efficiently with proper verification [6]. Transaction related algorithms are solved by the processing capacity of software on computers of miners. A block in a blockchain is encrypted which is basically a computer-intensive process. Any member of the group can submit a blockchain to the ledger, but the pertinent computation required to check about the fakeness, guarantee of submission is very intensive. The objective of blockchain function is consensus algorithm. Here we address two blockchain based consensus algorithm, one is known as proof-based and the other one is known as vote-based. Both the consensus algorithms are presented in the following subsections.

2.2 *Proof-Based Consensus Algorithm*

Here, we present a short description of the proof-based consensus algorithm (PBCA) as per reference [4]. Proof based consensus algorithms are as follows: (i) Proof of Work (PoW), (ii) Proof of Stake (PoS), (iii) Proof of Burn (PoB), (iv) Proof of Capacity (PoC) and (v) Proof of Elapsed Time (PoET). Proof based consensus algorithms are designed independently. However, a hybrid Proof based consensus algorithm can be possible to develop by combining any two of these Proof based consensus algorithm. The fundamental structure behind consensus algorithm is like this: among many other nodes, only one node performs sufficient proof. The same node will get its permission to append a new block to the chain and receive the award. PoW is originally proposed by Nakamoto [5].

2.2.1 PoW-Based Consensus Algorithm

According to Blockchain network, rising of confusion in network structure can be possible, if every node attempts to communicate their packet of information containing the validated and verified transactions. The said transaction can be duplicated in different packets and communicate to other packets if work related to broadcasting or communication is free but the ledger will be then meaningless. Whenever there is an addition of new packet/block in the chain, an agreement between all nodes pertaining to the information or message about the foreign added block will be established by PoW with adjusted difficulty for firing each node to solve the difficult task to generate a permission to update the system with a block which is new to the current chain. The first node performing sufficient proof or evidences will have this token to solve the task.

2.2.2 Variants Based on PoW

The drawbacks of PoW are observed in studies of many researchers. Drawbacks include some security issues and usage issues. With a view to this, several variants are developed to mitigate those drawbacks. Tromp [6] have proposed to replace the complicated work using Cuckoo hash function [7]. The said function allows data miners to apply some techniques to generate the permission to append the block easier. Basically, the Cuckoo hash function creates the relevant graph, wherein the vertices either in the hashed coordinate or replaced coordinate with values representing edges that have been deleted. One of the drawbacks of PoW proposed by many researches, is defined to be a problem known as the Double Spending Attack (DSA), a potential attack against improper usage of digital tokens [8]. In DSA, potential attacker would try to alter the flow path of a transaction verified by some other piece of information (nodes), with the help of another transaction which make the first one invalid in another fork. Attacker tries his best to make that transaction longer than the others.

2.3 Voting Based Consensus

Here in this case, execution of voting based consensus algorithm (VBCA) takes place by known adjustable nodes inside the verifying networkand as a result nodes can exchange the message very easily. The main difference between VBCA and PBCA is that in VBCA it is not possible to know accurately about joining and withdrawal of free nodes from the proved network. In VBCA, apart from maintaining the ledger, it is required to verify the transactions or blocks for all the nodes in the network. Prior to take a decision of whether to append their proposed blocks to their chain or not, an authenticate communication with others will take place.Execution of VBCA is very similar to standard methods to tolerate faults occurred in the distributed system.

2.4 Consensus with Paxos

Let us assume a collection of measurement systems that propose values such as velocity of flow of groundwater by various instruments. A consensus algorithm in this case ensures that a single measurement among many such similar measurements proposed is chosen. If no value is suitable then no measurement should be selected. If the velocity of groundwater flow has been chosen as the events to be aggregated or consensus, then aggregation of corresponding measurements by all means should be able to provide the desired point estimate of the same. Therefore, the safety of consensus from the point of failure can be either any of these tasks such as (i) a single (point) estimate proposed for its selection, (ii) selection of only a single value, and (iii) a measurement system that never learns. According to Morris [9], the replication approach of state machine for distributed computing is based on protocols followed in consensus algorithm. Literature survey also reveals the same [9]. The method to design an algorithm for a FT distributed system is defined as replication of state machine by Tromp [10]. Safe handling of all cases is ensured by Lamport et al. which is nothing but the principled approach. The protocols of Paxos are to generate a spectrum of policies within a cluster containing processors, messages that delay before learning the agreement, messages sent, types of failures and the activity of individual participants [11]. Consistency of Paxos algorithm says that there is no possibility of any guarantee about the progress in any asynchronous network by deterministic fault-tolerant consensus protocol by Schwarz [11].

If process is "replication of a file" or "an event of a database (distributed)", algorithm Paxos can be used because in that task durability is the main issue. The lifetime of the state can be as large as possible. Processes are attributed as nodes and three types of nodes exist in Paxos system as shown in Fig. 1. These processes are labeled as: (a) proposers, (b) acceptors and (c) learners. Activities of these processes are: proposers propose values that should be chosen by the consensus, acceptors form the consensus and accept values, and finally learners learn which value was chosen by each acceptor and therefore the consensus. If we the working function of a multitier

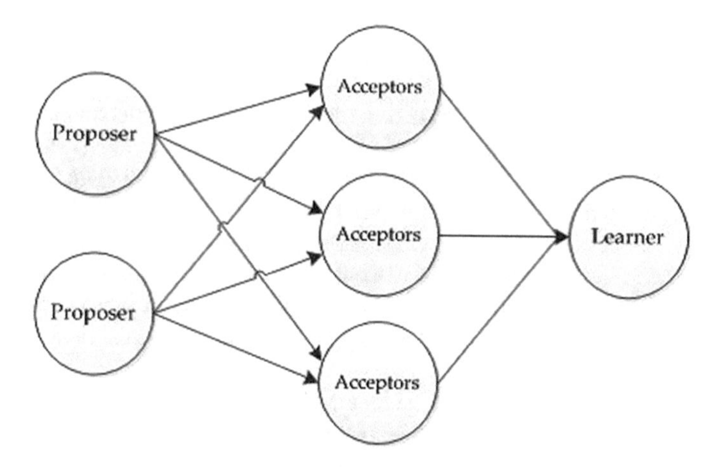

Fig. 1 Processes in Paxos

database, we can easily understand that the database client is likely to be a proposer as well as learner, in the sense that specific rules got fire to retrieve the information or any record and after invoking the same client learned the system's working principle. So, we can say that a process such as invoking of an event in a database may fill several roles at a same time.

Working rule of Paxos takes place through its various processors. Explicit assumptions and definitions simplify Paxos. Procedures to enhance its application are elsewhere found in the literature [12, 13]. Activities of processors of Paxos are grouped into (i) operation at arbitrary speed, (ii) notification of failures, (iii) processors having stable storage rejoining the protocol after failures and (iv) processors which are not lying or attempting to subvert the protocol [13].

Networking of the processors include the activities, viz., (a) sending messages to any other processor, (b) Asynchronous sending of messages and delivery of them by a long time arbitrarily, (c) recovery of messages that are lost, reordered, or duplicated and (d) delivery of messages without corruption. If N represents the number of processors and F represents processors that fail simultaneously, then PAXOS can make progress these N processors, in presence of F processors. The relation between N and F is $N = 2F + 1$ ($N > F$ always). Details of processors of PAXOS can be found elsewhere in [13]. Performance of protocol followed in operating PAXOS is assessed using optimization technique. A few of such optimizations events are: Accept and prepare. In accept event, messages are sent by acceptors to the designated learner [13]. Most often the leader and identified learner play their roles to perform by a designated processor [13]. In the event of prepare, a leader send its token to complete the requirement of quorum of acceptors [13]. Selection of exact value is not cared by acceptors. Respond to that token, messages take place to ensure about the selection of a single value in presence of failures. If acceptor receives later a token message as "Prepare or Accept", it can simply communicate the leader of the selected value [13].

2.5 Consensus with RAFT

Consensus algorithm Raft [14, 15] is designed to be easy to understand. Raft helps users to design and establish a variety of higher quality systems (consensus oriented) compared to the existing in the current market.In fault-tolerant (FT) distributed systems, consensus is a fundamental problem. It involves agreement on values of multiple servers. Once a decision on a value is reached, the evaluated decision is made final. Replicated state machines signify a commonsense attempt to build the FT systems and consensus typically arises in that domain [14, 15]. It is required to know that each server has a state machine and a log. One wants to make fault tolerant of the state machine, like a hash table.

Raft implements consensus by first identifying a distinguished leader and thereafter invokes complete responsibility to handle the replicated log. It accepts entries logged from clients, replicates them on other servers and send message to servers about the safety to their sate machines [16]. Possibility of failure or disconnection of a leader from the other servers results the election of a new leader. A coherent group named as Boltzmann machines is allowed by consensus algorithms. Novel features of Raft are: (a) Strong leadership, (b) Election of Leader and (c) Change of membership as tokenization. Operational details of these novel features of RAFT can be found elsewhere in [16].

As we know that consensus algorithms exhibit in replicated state machines, it is mandatory to describe about them. Service replication is a technique in which copies of the service are deployed in a set of servers instead of in just one. This technique is often used with two objectives: (a) to increase the performance of the system and (b) to increase the capacity or fault tolerance of the system. Replication can increase the service' performance and capacity since the replicas provide more resources to the service [16, 17]. Fault tolerance can be achieved with several replicas through the use of spatial redundancy to the system, making it continue to operate despite the failure of a fraction of these replicas [16, 17]. Consensus algorithm in general can be viewed in the event of replicated state machine [18]. In this case, computation of similar copies of the same state on a group of identified servers is executed by servers of state machines and the job can continue even if some of the servers are down. A large number of different FT problems in distributed systems are operated by replicated state machine.

3 Soft Computing Based Consensus Using EVICON

Consensus is an agreement on a single value or data point originated from various resources. If the information from these resources are imprecise or incomplete consensus is possible to achieve using soft computing technique where information or data nodes are attributed as fuzzy data. Each of these fuzzy nodes is valued by their membership function like triangular, trapezoidal and Gaussian. Value of membership function of a fuzzy data set itself originates from consensus of many people.

EVICON consists of several blocks of modules which communicate with respect to each other on the basis of evidence theory [19]. Consensus in EVICON is generated by arguments from several experts. Expert's opinions on a single event are aggregated either by mixing rule or Dempster rule of combination [19]. Here in this case, decision making for numerical values of parameters governing a physical system or any engineering model is taken based on consensus. The main objective of EVICON is to carry out the sensitivity and uncertainty analysis of any engineering system. The purpose of uncertainty analysis is to qualify the system in the sense that if the outcome of the system is highly uncertain then risk associated with the system from the point of its utilization in practice is very high and decision can be easily taken whether to use this system or not. So, to have the best quality system uncertainty analysis or risk informed decision making technique is essential. However, any engineering system being governed by a large number of parameters, uncertainty quantification of all parameters is not possible because physical as well as computational cost involved will be too high. So, prior to uncertainty quantification of the system, it is mandatory to execute the sensitivity analysis.

Sensitivity analysis (SA) of any physical or engineering model containing many parameters is generally executed to extract the most influential parameter of the model of interest [20–22]. Sensitivity analysis can be either deterministic or probabilistic. Industrial problems generally look for probabilistic method of sensitivity analysis where for each parameter of the model of interest, a probability distribution is required. However, to assign the probability distribution of a parameter, a large number of experiments are required which may not be possible. Repetition of experiments on a single parameter for many number of times increase sample size which is essential to construct the pertinent probability distribution of that parameter. Not only that, suitable experiments for many of the parameters may not be possible to carry out at all. In that case, probability distribution of those parameters cannot be possible and accordingly those parameters are imprecise. Therefore, probabilistic method of sensitivity analysis fails very often. Accordingly researchers look for an alternate method of sensitivity analysis. In order to fulfil the demand of researchers, possibilistic method of sensitivity analysis is addressed here. EVICON provides the consensus methodology to devise possibilistic method of sensitivity analysis.

Therefore, it is mandatory, to do the sensitivity analysis prior to quantification of uncertainty of the model of interest. For example, in the field of fluid mechanics, model describing the system behavior by advection diffusion reaction mechanisms, measured concentration of contaminant presence in any aquatic stream (river, estuary, coastal or lake) consists of error and the research question is whether the error is acceptable? Answer to this kind of query needs uncertainty quantification of the representative model. Traditional method of Monte Carlo simulation cannot be used for carrying out the SA due to small sample size of the parameters of interest [23].

The goal of this section is to present the sensitivity analysis for models having uncertain parameters which are imprecise or possibilistic, due to their lack of knowledge or insufficient measurements. In the framework of possibility theory, sensitivity analysis is based on the possibility distribution of the model parameters [23]. Possibility distribution is generally addressed in terms of a fuzzy number [24]. Membership grade of a fuzzy number being subjective, the corresponding possibility distribution of the model parameters has been addressed in terms of Dempster-Shafer (D-S) structure [25]. D-S structure is defined over a power set consisting of an interval valued set (focal element) with a basic probability assignment (BPA) or basic mass assignment (BMA), a number $\in [0, 1]$. We propose a new methodology of SA based on consensus of experts. Consensus is built on the basis of evidence theory [25]. SA is demonstrated using standard one dimension contaminant transport problem (Advection-Diffusion Model). Two parametersof the said model, i.e., the longitudinal dispersivity (α L) and the velocity of flow of the river (u) Model are taken into account as uncertain and their uncertainty is due to their imprecision which justifies the utility of EVICON. The target is to elect the most sensitive parameter (leader) between α and u.

3.1 Dempster-Shafer Evidence Theory

Dempster-Shafer theory (DST) is based on twonon-additive fuzzy measures named as belief and plausibility [25–27]. A consensus operator is formulated in EVICON and this operator is nothing but a triplet consisting <belief, disbelief, uncertainty>. In fact plausibility is the sum total of disbelief and uncertainty. DST is a theory of plausible reasoning since it addresses the fundamental operation of belief, disbelief and uncertainty which is further based on the combination rule of degrees of truth (belief) according to different evidences. The BPA of D-S structure is defined as probability like assignment, $m : 2^{\theta} \to [1]$ to the focal element (an interval) such that

$$m(0) = 0 \tag{1}$$

$$\sum_{A \in \theta} m(A) = 1 \tag{2}$$

where θ represents a frame of discernment [27] and the number $m(A)$ signifies the BPA.

3.2 Belief and Plausibility Function

If X represents the universal set, belief (which is basically a measure) function of DST is defined as a function, $BL : P(X) \to [0, 1]$ such that $BL(\Phi) = 0$ and $BL(X) = 1$. Let m be a given BPA. We now have a function $BL : 2^{\theta} \to [0, 1]$, called a belief function over θ if and only if

$$BL(A) = \sum_{B \subseteq A} m(B) \tag{3}$$

On the other hand, the plausibility function, denoted by PLS, is represented as a function $PLS : 2^{\theta} \rightarrow [0, 1]$ such that

$$PLS(A) = 1 - BL(\bar{A}) \tag{4}$$

The plausibility function $PLS(A)$ can be expressed in terms of the BPA, m of BL as

$$PLS(A) = 1 - BL(\bar{A}) = \sum_{B \subseteq \theta} m(B) \sum_{B \subseteq A} m(B) = \sum_{B \cap A = \emptyset} m(B) \tag{5}$$

If the intervals B_i are nested according to the Eq. (6):

$$B_i \subseteq B_{i+1}\, for\ i = 1, 2, \ldots, n - 1 \tag{6}$$

Then they can be given a basic assignment $m = 1/n$. because the B_i intervals are equally credible. The belief and *plausibility* measures [27] can be computed as follows:

$$BL(B_i) = \sum_{all B_j \subseteq B_i} m(B_j) \tag{7}$$

$$PLS(B_i) = \sum_{all B_j \cap B_i} m(B_j) \tag{8}$$

Aggregation of opinions resourced from many experts are combined using Dempster rule of combination. Basically, in the evidence theory, evidences are pooled from more than one resources for aggregation. Mathematical representation of such combination (D-S rule of combination) is given by:

$$m_1 \oplus m_2(C) = \begin{cases} 0, & \text{if } C = \emptyset \\ \frac{\sum_{A \cap B = C} m_1(A) m_2(B)}{1 - \sum_{A \cap B = \emptyset} m_1(A) m_2(B)}, & \text{otherwise} \end{cases} \tag{9}$$

where the symbols have usual significances.

3.3 *Sensitivity Analysis Using Consensus*

Our algorithm to carry out sensitivity analysis of an engineering or physical system is based on evidence theory based consensus known as EVICON. Identification of most influential or most sensitive parameters of a system or any model using

Monte Carlo simulation is carried out on the basis of correlation coefficients between the model output and the input parameters of the model. Higher the correlation coefficient of the input parameter with the model output, larger is the sensitivity of the parameter of interest. On the contrary, similar task in the possibilistic domain using EVICON is carried out in terms of the average width of the area (p-box) bounded by belief-plausibility of the model output generated successively by pinching one of the parameters. Pinch method is basically appliedto identify the sensitivity of imprecise parameters of the model of interest. In the pinch method, parameters of the model are pinched (fixed) and corresponding belief and plausibility contour of the model output is constructed. Belief and plausibility contour of the model output is finally constructed without pinching of all the parameters. Finally the average width of the belief-plausibility contour of the pinch method and that without pinching of all parameters is estimated. The average width is the cumulative width of all focal elements weighted by their mass. The smallest average width of the belief-plausibility contour for specified pinched parameter is proposed as an indicator for the most sensitive parameter of the model. Non-specificity measure of uncertainty [28] is also computed for SA. Smallest value of the non-specificity with pinching of the specified parameter indicates that the specified parameter is most sensitive.

3.4 Case Study: Contaminant Transport

Contaminant from any chemical industry is discharged as the liquid effluent into surface water (river) [29]. Flow velocity of the river aids in transporting this contaminant in the direction of flow of the river water. Diffusion of the contaminant also takes place along with its transport. Hence, longitudinal dispersivity and the velocity are the two parameters of the governing model (advection-diffusion model) considered for evaluation of SA. The model computes the concentration of the contaminant at any time and at any downstream distance. The uncertainty distribution of the parameters is evaluated using EVICON. Consensus of five experts is utilized to address the uncertainty of both these imprecise (possibilistic) parameters. The governing equation of one dimension contaminant transport is nothing but advection-diffusion equation, which is numerically solved using specific initial and boundary conditions. Forward time finite difference numerical method with imprecise parameters is used to have numerical simulation of contaminant concentration at any temporal and spatial location. Conflicts between two or more experts are evaluated using D-S rule of combination. One can also weight mixing rule of combination or deep learning techniques. Outcome of the parametric uncertainty using EVICON is shown in Figs. 2 and 3. Initial concentration of the contaminant C_0 is assumed as 100 mg/l for simulation. Concentration of the contaminant $C(x, t)$ is computed at time 400 days and at spatial coordinate (downstream distance) of 1220 m. Consensus of experts'opinion, suggests uncertainty of both the parameters as uniform distribution and accordingly within their specified range simulation is carried out using Monte Carlo technique. Belief- plausibility contour of the concentration of the contaminant without pinch

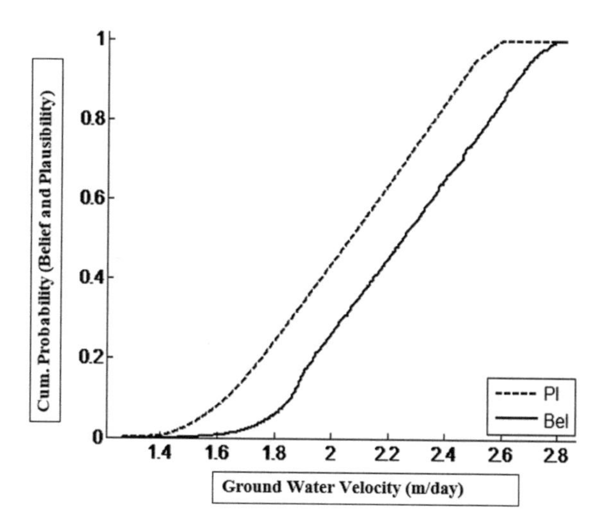

Fig. 2 Belief-plausibility contour of the longitudinal dispersivity

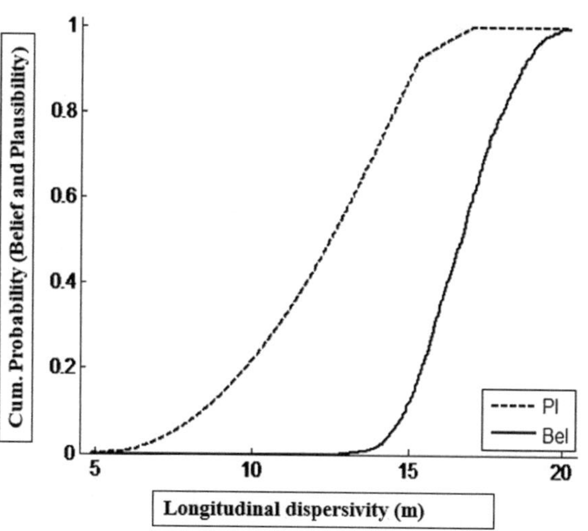

Fig. 3 Belief-plausibility contourof the velocity

and with pinch (velocity and the longitudinal dispersivity respectively) is computed at $x = 1220\,\text{m},\ t = 400$ days. Results are shown in Figs. 4, 5 and 6 respectively.

SA of the uncertain parameters of the model is notified by computing the average width of Belief-Plausibility contour (BPC) of the concentration of the contaminant under the condition of (a) without pinching of parameters, (b) pinching the velocity of flow and (c) pinching the longitudinal dispersivity. Average width of all these BPC is estimated and tabulated in Table 1. Smallest average width of the BPC is considered as the sensitivity index and the parameter which is considered as pinched (no uncertainty) for that BPC is indicated as the most sensitive parameter. We have also measured the sensitivity using the non-specificity uncertainty measure of the model output. Non-specificity of the concentration for all the three conditions are

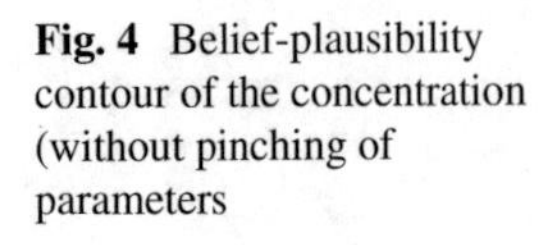

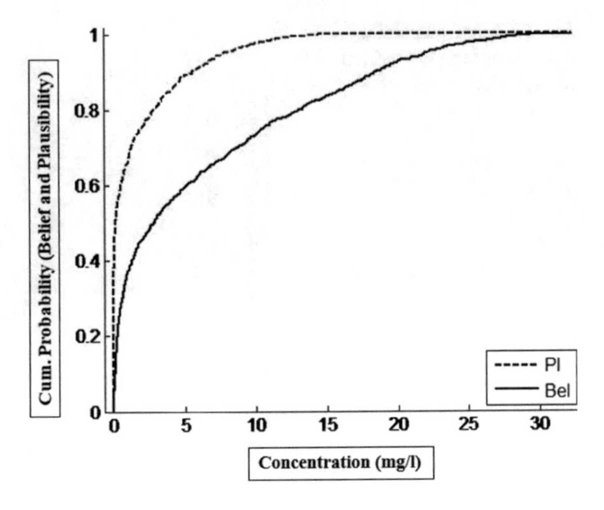

Fig. 4 Belief-plausibility contour of the concentration (without pinching of parameters

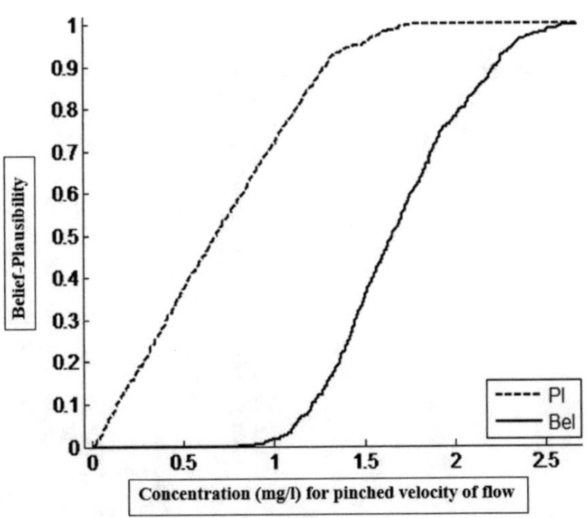

Fig. 5 Belief–plausibility contour of the concentration (flow velocity is pinched)

tabulated in Table 1. It can be easily seen from Table 1, that the velocity of flow is the most sensitive parameter as the average width of the BPC of the contaminant concentration for pinching the velocity of flow is smallest as well as the corresponding non-specificity value is the smallest. Therefore, EVICON based consensus identifies the parameter 'velocity of flow' is most sensitive.

4 Conclusion

A short review of available consensus algorithms is presented. Consensus algorithms are often designed with correctness and efficiency. Consensus algorithm is presented

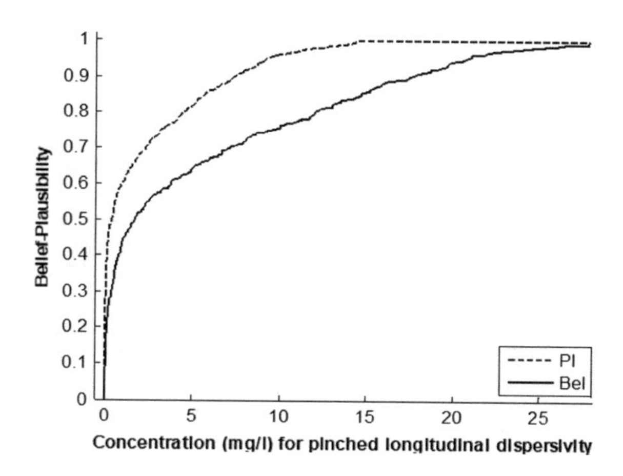

Fig. 6 Belief-plausibility contour of the concentration (longitudinal dispersivity is pinched)

Table 1 Sensitivity indicator

Pinched status	Average width of belief-plausibility contour	Non-specificity for the measures of the concentration
No pinch	4.767	1.8978
Velocity is pinched	0.973	0.9748
Longitudinal dispersivity is pinched	3.407	1.6045

in terms of evidence theory, where belief and plausibility measures are exploited. Deep understanding of the algorithm is essential otherwise it is difficult to retain its desirable properties in the implementation of consensus algorithm. A new method for sensitivity analysis of any model is proposed. New method is based on Dempster-Shafer evidence theory. Model consisting of non-probabilistic or imprecise uncertain parameter can be used for sensitivity analysis. Expert's opinion is utilized for collecting the uncertain information of the model parameters as they are insufficient usually. The average width of the belief-plausibility (measures of evidence theory) contour and the non-specificity measure of uncertainty are presented as the sensitivity index. Evidence theory shows promising to handle sensitivity analysis with imprecise parameters of a model. Advection-Diffusion model (one dimension) is chosen for illustration of the new method of sensitivity analysis.

Acknowledgements Authors are thankful to Guru Gobind Singh Indraprastha University and Bhabha Atomic Research Centre for research facility.

References

1. Pease, M., Shostak, R., Lamport, L.: Reaching agreement in the presence of faults. J. Assoc. Comput. Mach. **27**(2) (1980)
2. Lamport, L.: The implementation of reliable distributed multiprocess systems. Comput. Netw. **2**, 95–114 (1978)
3. Nguyen, G.T., Kim, K.: A Survey about consensus algorithms used in blockchain. J. Inf. Process. Syst. **14**(3), 101–128 (2018). https://doi.org/10.3745/JIPS.01.0024
4. Google Press Center: Fun Facts. Archived from the original on 2001-07-15
5. Lamport, L.: The part-time parliament. ACM Trans. Comput. Syst. **16**(2), 133–169 (1998)
6. Schneider, F.: Implementing fault-tolerant services using the state machine approach: a tutorial (PDF). ACM Comput. Surv. **22**(4). 299-319 (1990). CiteSeerX 10.1.1.69.1536. https://doi.org/10.1145/98163.98167 Leslie
7. Lamport, L.: Time, clocks, and the ordering of events in a distributed system. Commun. ACM **21**(7), 558–565 (1978)
8. Lamport, L.: Paxos made simple. ACM SIGACT News (Distributed computing Column) **32**(4), 51–58, 121
9. Morris, D.Z.: Leaderless, blockchain-based venture capital fund raises 100 million and counting. Fortune (2016). Archived from the original on 21 May 2016. Retrieved 23 May 2016
10. Tromp, J.: Cuckoo cycle: a memory-hard proof of work system (2015). https://eprint.iarc.org/2014/059.pdf
11. Schwarz, K.: Cuckoo hashing. http://web.stanford.edu/class/cs166/lectures/13/small13.pdf
12. Bitcoinwiki: Irreversible transactions. https://en.bitcoin.it/wiki/IrreversibleTransactions
13. Lamport, L.; Massa, M.: Cheap Paxos. In: Proceedings of the International Conference on Dependable Systems and Networks (DSN 2004) (2004)
14. Liskov, B., Cowling, J.: Viewstamped replication revisited. Technical Report: MIT-CSAIL-TR-2012-021, MIT (2012)
15. OKI, B.M., Liskov, B.H.: Viewstamped replication: A new primary copy method to support highly-available distributed systems. In: Proceedings of ACM Symposium on Principles of Distributed Computing, PODC'88, pp. 8–17. ACM (1988)
16. Ghemawat, S., Gobioff, H., Leung, S.T.: The google file system. In: Proceedings of ACM Symposium on Operating Systems Principles, SOSP'03, pp. 29–43. ACM (2003)
17. Shvachko, K., Kuang, H., Radia, S., Chansler, R.: The hadoop distributed file system. In: Proceedings of Symposium on Mass Storage Systems and Technologies, MSST'10, pp. 1–10. IEEE Computer Society (2010)
18. Ousterhout, J., Agrawal, P., Erickson, D., Kozyrakis, C., Leverich, J., Mazieres, D., Mitra, S., Narayanan, A., Ongaro, D., Parulkar, G., Rosenblum, M., Rumble, S.M., Stratmann, E., Stutsman, R.: The case for RAM Cloud. Commun. ACM **54**, 121–130 (2011)
19. Dempster, A.P.: Upper and lower probabilities induced by a multiple valued mapping. Ann. Math. Stat. **38**, 325–339 (1967)
20. Saltelli, A., Andres, T.H., Homma, T.: Sensitivity analysis of model output: an investigation of new techniques. Int. J. Comput. Stat. Data Anal. **15**, 211–238 (1993)
21. Helton, J.C.: Uncertainty and sensitivity analysis techniques for use in performance assessment for radioactive waste disposal. Int. J. Reliab. En. Syst. Safety. **42**, 327–367 (1993)
22. Saltelli, A., Chan, K., Scott, E.M.: Sensitivity analysis. In: Wiley Series in Probability and Statistics. Wiley (2000)
23. Datta, D.: Statistics of Monte Carlo methods used in radiation transport calculation, applications of Monte Carlo methods. In: Kushwaha, H.S. (ed.) Nuclear Science and Engineering. Bhabha Atomic Research Centre, Trombay, Mumbai (2009). ISBN 978-81-8372-047-2
24. Datta, D., Kushwaha, H.S.: In: Kushwaha, H.S. (ed.) Fundamental Statistics for Uncertainty Analysis, Uncertainty Modeling and Analysis, pp. 1–48. Bhabha Atomic Research Centre (2009). ISBN 978-81-907216-0-8
25. Shafer, G.: A Mathematical Theory of Evidence. University Press, Princeton (1976)

26. Dubois, D., Nguyen H.T., Prade, H.: Possibility Theory, Probability and Fuzzy Sets: Misunderstandings, Bridges and Gaps, Fundamentals of Fuzzy Sets, pp. 343–438. Kluwer Academic Publishers, Boston (2000)
27. Klir, G.J., Wierman, M.J.: Uncertainty-Based Information. Springer (1998)
28. Yager, R.R.: Entropy and specificity in a mathematical theory of evidence. Int. J. Gen. Syst. **9**, 249–260 (1983)
29. International Atomic Energy Agency: Hydrological Dispersion of Radionuclide Material in Relation to Nuclear Power Plant Siting, Safety Series No. 50-SG-S6, IAEA, Vienna (1985)

Smart Contracts-Enabled Simulation for Hyperconnected Logistics

Quentin Betti, Benoit Montreuil, Raphaël Khoury and Sylvain Hallé

Abstract The combination of the Internet of Things and blockchain-based technologies represents a real opportunity for supply chain and logistics protagonist-s, who need more dynamic, trustworthy and transparent tracking systems in order to improve their efficiency and strengthen customer confidence. In parallel, hyperconnected logistics promise more efficient and sustainable goods handling and delivery. This chapter shows how the Ethereum blockchain and smart contracts can be used to implement a shareable and secured tracking system for hyperconnected logistics. A simulation using the well-known AnyLogic software tool provides insights on the monitoring of properties depicting shipment lifecycle constraints through a stream of blockchain log events processed by BeepBeep 3, an open source stream processing engine.

1 Introduction

Supply chain logistics typically involves, among other things, managing the deployment of physical goods over geographical markets and their delivery over both long and short distances. The Physical Internet [1] notably aims at transforming goods handling, storage and transportation towards a most efficient supply chain management in terms of delivery time, cost, as well as social and environmental impacts. Physical Internet enabled *hyperconnected* logistics has emerged as a radical redesign of classical supply chain networks, where each parcel to be delivered is treated as a

Q. Betti · R. Khoury · S. Hallé (✉)
Université du Québec à Chicoutimi, Québec, Canada
e-mail: shalle@acm.org

Q. Betti
e-mail: quentin.betti1@uqac.ca

R. Khoury
e-mail: Raphael_Khoury@uqac.ca

B. Montreuil
Georgia Institute of Technology, Georgia, USA
e-mail: benoit.montreuil@isye.gatech.edu

M. A. Khan et al. (eds.), *Decentralised Internet of Things*, Studies in Big Data 71,
https://doi.org/10.1007/978-3-030-38677-1_6

physical form of “packet” that can be routed across hubs according to its own lifecycle. Recently, the concept of hyperconnected logistics has been coupled with the concept of city logistics to radically transform the way physical goods flow through urban environments, through a vision of hyperconnected city logistics [2]. These new paradigms give rise to a totally distributed multi-party system where each parcel physically carries a fragment of the supply chain’s state with it as it moves around the globe. In such a setting, making sure that the supply chain remains in a consistent state and detecting various kinds of errors becomes a challenging task.

The triple emergence of the Physical Internet, the Internet of Things (IoT) and blockchain technologies [3–10] holds the promise of a next generation supply chain that is both more efficient as well as free from the requirement that a central authority be entrusted by all parties with their confidential data. Extending a recently published work of the authors [11], the first aim of this chapter is to show concretely how one can leverage blockchain technologies to create a distributed, ownerless, and secure log of all events related to an entire supply chain. This is done by integrating a blockchain-based backend to an existing simulation [12] designed with AnyLogic [13]—a widespread supply chain simulation software—where transactions related to parcels are stored in a private Ethereum blockchain through smart contracts.

An appealing advantage of such a setup is the capacity to query this blockchain for a variety of properties of interest that guarantee the correct and efficient operation of the supply chain. The second goal of this chapter, thus, is to define several kinds of such properties and implement them through the use of the BeepBeep stream monitor [14]. Firstly, we define what we call *correctness* properties. A violation of these properties typically indicates a malfunction or malicious tampering of the supply chain; for example, a property stating that every manipulation performed on a parcel takes it closer to its destination, or that no package is left unattended for too long. Then, we introduce two other types of properties, previously less studied in the literature. First, *analytical* properties are non-Boolean properties that provide actionable information about the status of the supply chain, such as the statistical distribution of shipping time between two endpoints. Second, *self-correlated* properties are parameterized queries that compare a trend computed on a recent piece of the log to its own past. For instance, a property stating that the delivery time for a parcel must not exceed the historical average of the shipping time on the same route by a factor of more than k is called a self-correlated property. Modelling these properties requires a broad mix of modelling notations (from Moore machines to slicing and set manipulations) as well as data types for events (tuples, scalars, histograms, among others).

Finally, the results of such blockchain-based supply chain scenario as well as the property monitoring performance are discussed. They show, on one hand, that blockchain technology is highly valuable in such context and that current platforms almost have all the requirements for these applications, despite their youth. On the other hand, experimental evaluations point out that a monitor for property verification can handle the typical workload of thousands of parcels in real time, thus meeting current supply chain scenarios’ expectations. This chapter contributions are therefore twofold: first, it introduces a new concrete use case making use of blockchains;

second, it provides a set of new properties to be computed on this blockchain, which pushes the boundaries of the state of the art regarding their expressiveness.

The remainder of this chapter is organized as follows: Sect. 2 gives some background on the Physical Internet as well as hyperconnected logistics, and presents a simulation designed with AnyLogic illustrating those principles which will serve as basis for the rest of this chapter. Next, in Sect. 3, we highlight the expected benefits of blockchain technology for supply chain, by laying its technical foundations and listing several existing applications to this domain, and we detail the design and implementation of the integration of a blockchain backend to the AnyLogic simulation mentioned previously. Then, Sect. 4 states a set of interesting properties and shows how verification tools, such as BeepBeep, can be used to monitor these by capturing blockchain transactions. Finally, results of both blockchain integration and property monitoring are discussed in Sect. 5.

2 The Physical Internet and Hyperconnected Logistics

The hyperconnected logistics paradigm, in line with the Physical Internet [1], aims at transforming the current supply chain and logistic industries in order to make it more efficient in terms of cost and sustainability [2]. It also has the goal of improving goods handling, routing and delivery speed, especially in cities. This is particularly achieved by evolving away from a hub-and-spoke architecture towards so-called *hyperconnected* networks [15]. In this section, we shall first present the Physical Internet fundamentals, then define the parcel management framework in hyperconnected logistics, and finally outline an existing simulation of such concept.

2.1 Physical Internet Principles

The advent of the Physical Internet, introduced by Montreuil [1] is about improving by an order of magnitude the efficiency, capability and sustainability of fulfilling society's demand for physical object services. This includes the way we move, deploy, realize, supply, design and use physical objects. The Physical Internet has been formally defined by Montreuil [16] as a global hyperconnected logistics system enabling seamless open asset sharing and flow consolidation through standardized encapsulation, modularization, protocols and interfaces. The system is said to be hyperconnected as its components and actors are intensely interconnected on multiple layers, ultimately anytime, anywhere. The interconnectivity layers of the Physical Internet include the digital, physical, operational, business, legal and personal layers. According to [17], the key building blocks for full implementation and large-scale adaption and worldwide exploitation of the Physical Internet are (1) a unified set of standard modular logistics containers; (2) containerized logistics equipment and technology; (3) standard logistics protocols; (4) certified open logistic facilities and ways; (5) global logistics monitoring system; (6) open logistic decisional and transactional platforms, smart data-drive analytics, optimization and simulation; and (7) certified open logistics service providers. Ultimately, [1] envisions that the Digital Internet, the Internet of Things, the Physical Internet and the Energy Internet will develop in the set of hyperconnected infrastructures that are to be the pillars of economy and society, enhancing each other's capabilities, efficiency and sustainability, and all drastically reducing marginal costs of exploiting them (as advocated in [18]).

2.2 The Hyperconnected Logistics Paradigm

In the hyperconnected model (Fig. 1), the entire world can be split at the smallest scale into *unit zones*, whose size depends on expected demand density. Adjacent unit zones are grouped into local *cells*, which in turn are gathered into *areas*, which form *regions*. Simultaneously, several hub networks are defined to link these different layers: *access hubs* link unit zones together; *local hubs* link local cells, and *gateway hubs* link areas. Different hub levels may exist inside the same physical entity (e.g., a local hub might also be an access hub), thus allowing interactions between the different layers.

The concept of hyperconnected logistics introduces several disruptions with respect to traditional supply chain logistics. For example, a container that must be delivered across a long distance between points A and B will typically be put on a truck that will ride this route from start to end. In hyperconnected logistics, on the contrary, this same container will repeatedly jump between smaller hubs and be transferred from one truck to the next until reaching its destination. Upon switching to a lower layer of the hyperconnected plane, the container's contents may even be broken down into smaller parcels, with each parcel following a different sequence of hubs towards destinations within the unit zone.

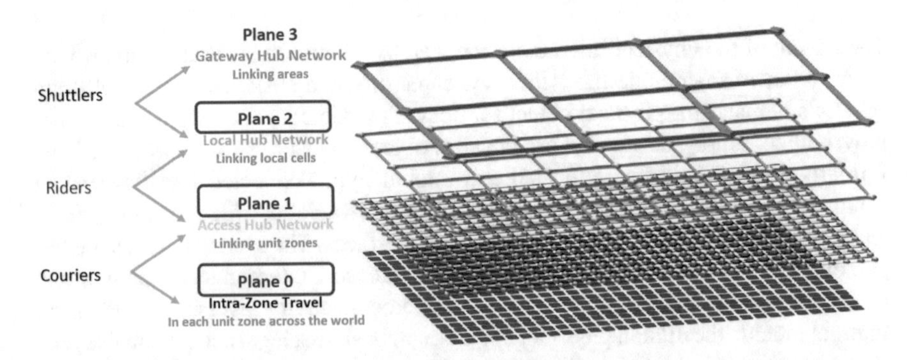

Fig. 1 Urban parcel logistic web from Montreuil et al.[15]

2.3 *An AnyLogic Simulation for Hyperconnected Logistics*

To illustrate the potential of such architecture on supply chain efficiency, a simulation model of hyperconnected urban logistics was developed by Kaboudvand et al.with the AnyLogic software and presented at the last IISE Annual Conference [12].

In this simulation, the authors consider the following simple model of a hyperconnected "megacity", using hyperconnected logistics concepts defined in 2.2. The city itself is modeled as a rectangular map and actually limited to one area (see Fig. 2); this area is itself composed of four local cells. Shipments to be delivered pop up across the city, and both transporters and deliverers participate in shipping them to their final destination inside the city. Each local cell is divided in nine unit zones, whose brightness reflects their demand density: the darker a unit zone's color is, the higher the probability that the zone will be the origin or the destination of a shipment request. Access hubs are located at each vertex of a unit zone and are shared with their adjacent unit zones. Local hubs are located at each vertex of a local cell and are shared by their adjacent local cells. In this scenario, local hubs are also access hubs.

Three kinds of carriers are defined in this model: *couriers*, *riders* and *shuttlers*. Couriers take care of last-mile deliveries within their assigned unit zone, doing so walking, biking, or yet riding a scooter or light electrical vehicle. Riders and shuttlers use faster vehicles (usually electrical or natural gas) as they cover longer distances. Riders are assigned to a local cell, moving goods between unit zones and/or local hubs within the cell. Shuttlers are assigned to an area, moving goods between local

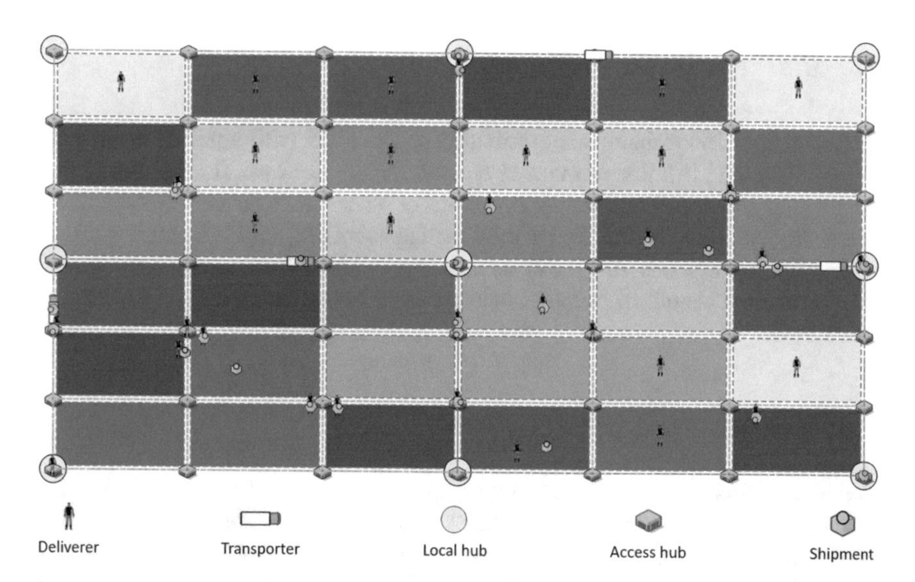

Fig. 2 Representation of the megacity and its components [12]

hubs and/or gateway hubs within the area. Shuttlers preferably use faster roadways than riders, similarly as riders preferably use faster roadways than couriers. For the sake of simplicity, here riders and shuttlers are gathered in a group of carriers called *transporters*.

3 A Blockchain-Based Tracking System

A blockchain can be seen as an immutable database where each entry is authenticated. Moreover, some of the blockchain platforms allow users to deploy *smart contracts*, pieces of code run by each node in the network. In this section, we first explore the reasons why such characteristics are particularly interesting for supply chain and logistics fields. This is done by briefly explaining the concepts of blockchains and smart contracts as well as listing the existing applications and potential benefits to these subjects.

Then, based on a previous publication by the authors [11], we proceed with designing a concrete tracking system based on an Ethereum blockchain backend that can be applied for hyperconnected logistics, and integrate it to the AnyLogic simulation presented in Sect. 2.3. This implementation provides a tracking system where actions performed on shipments by operators in the simulation must be stored using a blockchain system, instead of regular databases or log files.

3.1 Blockchains for the Supply Chain

The concept of blockchain was first introduced in 2009 with a now famous proof of work: Bitcoin [19], first of many cryptocurrencies. However, its application field has not stop growing ever since. Applications of blockchains include, among others, healthcare [20], biometrics [21], business management [22], and IoT security [23]. In this section, we first describe the basic concepts of blockchains and smart contracts and explain their benefits for supply chain management by listing several applications to this field.

3.1.1 Blockchain and Smart Contracts

Trivially, a blockchain is a data structure composed of *blocks* that contain *transactions*: it is *chained* because each block references its predecessor (Fig. 3). In a blockchain network, each node holds a copy of the blockchain, is able to read and send transactions, and new blocks are appended upon reception. Within a block, each transaction is signed by its author making impersonation impossible. Moreover, modifications or deletions of previously stored blocks and transactions are prevented using hashes: any illegitimate manipulation would be identified and

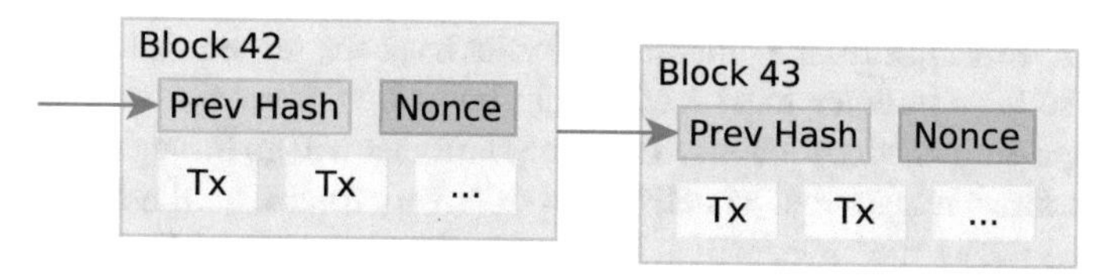

Fig. 3 General blockchain structure, adapted from Nakamoto [19]

rejected by the nodes. For these reasons, blockchains are considered to be shared, open, secure and trustworthy ledgers. However, there is no literal *trust*. Instead, blocks and transactions hold evidence of their authenticity based on hashes and a consensus (e.g., Proof-of-Work, Proof-of-Stake and Proof-of-Authority) the nodes agreed on beforehand. Upon reception of a new block, the nodes will compute these proofs, making sure the block and the transactions it contains are valid.

These features are especially relevant for *cryptocurrencies* such as Bitcoin, Ether, or Litecoin. In this scenario, the blockchain is used as a record of financial transactions between users, who are provided coins by buying them with common currencies, by mining (i.e., participating in new block generation according to a consensus) or just by receiving them from other users. Users are then free to spend these coins by sending transactions to others.

However, blockchains soon allowed the deployment and use of *smart contracts*. Theorized by Nick Szabo in the late 1990s [24, 25], a smart contract is "a set of promises, specified in digital form, including protocols within which the parties perform on these promises" [24]. As implemented in the Ethereum blockchain platform—the first to provide full smart contracts handling—a smart contract is just computer code that can be called by transactions sent to its address. Newly received corresponding transactions will then trigger the code to be executed on each node of the blockchain network. Any user can write its own smart contracts with the Solidity language [26], deploy them to the blockchain and execute their code, given he has sufficient funds.

An Ethereum smart contract is therefore a very complete entity: it has an account, can handle transactions and execute related code, has its own memory storage, and can even trigger what we call log events. Any application connected to the smart contract may listen to these log events, allowing a smart contract to be used like a regular web API. In that way, smart contracts are very powerful.

3.1.2 Benefits and Applications for Supply Chain

In the fields of supply chain and logistics, numerous actors have to collaborate in order to fulfill a demand. Here is a simple example: a buyer makes an online request to a seller for a specific item. Since the seller might not have its own website to merchandise its products, the order will likely be placed through an intermediate retailer or marketplace such as eBay or Amazon. Once the payment is done, the marketplace must inform the seller that an order has been placed. After confirmation from the seller, the latter must retrieve the buyer's address and ship the product through a posting service provider. Then, the shipment will go through different hubs and warehouses, transported by deliverers or delivery trucks, until it reaches its final destination, where the buyer will acknowledge the reception of the shipment. Note that, in this scenario, we already have six different kinds of actors: the buyer, the seller, the intermediate marketplace, the posting service provider, and the deliverers/delivery truck drivers.

Obviously, this example is considerably simplified, yet it would not be easy to identify the reasons for, say, a shipment loss or a delivery delay. Did the final deliverer hand over the packet to the buyer or keep it for himself? The buyer might be lying, got the product and is trying to get another one for free. The post service may have had a failure in its process and forgot the shipment, or maybe it was delivered at a wrong address. Thus, even in this little example, identifying failures in goods transport and find the right solutions would be highly complex and time-consuming without a proper trustworthy tracking system. Imagine then if we had a complete supply chain scenario, where parts of the product come from different suppliers and are carried through several shipping service providers across numerous countries, before being assembled and transported to actual retailers where, at last, they will be bought or delivered to costumers.

Moreover, since each actor uses most likely its own internal tracking system, when one of them is processing the shipment the others are completely left in the dark: they have to rely blindly on the data provided by the actor handling the parcel. This leaves the door open for thefts, counterfeits, and tampering of product origin, which are critical matters for companies and consumers, especially in the food, pharmaceutical and luxury industries. To limit these risks, third party intermediaries are often needed, but procedures can be slow, costly and yet not entirely risk-free, since corruption is still possible [27].

Circumventing these issues, blockchain can be used to enable a transparent and trustworthy traceability system [28–30]. Indeed, as we mentioned earlier, a blockchain is an open, decentralized, and immutable record where each transaction is authenticated. By adopting this technology in a shared tracking system, actors would have a way greater confidence in the data it contains since a malicious actor or attacker would not be able to alter them or impersonate others. For example, in [7] the authors pointed out the benefits of such technology with an application to supply chain management in aircraft production. In 2016, BigchainDB, a blockchain-based decentralized database, was introduced to tackle the scalability issues of regular blockchains [31]. Then, Tian showed how BigchainDB could be used in food supply chain traceability systems with the help of the IoT [3]. Toyoda et al.illustrated the potential of Ethereum blockchain and smart contracts to enforce an anti-counterfeit system in post supply chain [9], relying on RFID tags and SGTIN-96, a standard for global trade item numbers. Besides, as the use of the IoT is more and more expected in supply chain [8, 10, 32], blockchain can also be deployed to increase IoT security in this context [23]. Actually, many works describe the interests and implementations of blockchain technology in supply chain management, and Wang et al.recently realized a comprehensive survey of such publications and current industrial developments [30]. In this chapter however, we focus on the adoption of blockchain and smart contracts in common logistics simulation tools, and the monitoring of parcels' lifecycle properties.

3.2 *Ethereum Blockchain and Smart Contracts Setup*

In this section, we detail the design of the blockchain backend, which will be integrated in Sect. 3.3 into the AnyLogic simulation previously presented in Sect. 2.3. In this scenario, the processing of shipments inside a hub is not taken into account, meaning that the only actions being tracked are parcel deliveries and pickups by deliverers and transporters. These two entities are the *agents* of our simulation.

To enable such system, we simulate an Ethereum blockchain network. Although Ethereum clients are not the most optimized or fastest, the Ethereum platform was the first to implement smart contracts deployment and usage in a cryptocurrency network. Moreover, it is open-source, and benefits from a huge community, making it easy to find documentation on how to deal with smart contracts.

3.2.1 A Private Blockchain Network

In this simulation, our blockchain network is private: this means that several Ethereum nodes are running and communicating with each other, and that the blockchain they deal with is completely independent from other networks such as the main net[1] (where real cryptocurrency is used) and test nets, like Kovan[2] or Rinkeby[3] (which are mainly used by developers to test their decentralized applications). Doing so allows us to personalize the blockchain network configuration. For example, our network uses a Proof-of-Authority (PoA) consensus to generate blocks of submitted transactions, and there is no minimal transaction gas price in order for a transaction to be validated and inserted in a block, which is not the case with other networks. Furthermore, this also permits us to run all the nodes locally, thus removing the transaction submission/block reception delays due to packets' transport through the Internet. Finally, running our own private network makes it easy to allocate initial funds to the Ethereum accounts of the agents in the simulation.

In order to simulate a full-scaled network, we choose to assign a full Ethereum client to each agent of the AnyLogic simulation. This means that each deliverer and transporter receives all the transactions of the network, validates them on reception, stores them, and thus keeps the whole blockchain locally. Additionally, we run sealer nodes, i.e., network nodes that are responsible for transaction verification and block

[1] https://etherscan.io/.

[2] https://kovan.etherscan.io/.

[3] https://rinkeby.etherscan.io/.

generation when using a Proof-of-Authority consensus. Sealers are not assigned to any agent, they can be seen as servers made available by various companies and institutions to ensure the proper functioning of the blockchain network. Even though the mechanism of the PoA consensus itself implies that the majority of sealers must be trusted, it is a way lighter mechanism than Proof-of-Work in terms of computation, which is the reason of our choice since we already run several nodes on the same computer.

3.2.2 Solidity Smart Contracts

The Ethereum platform allows the implementation of smart contracts using the Solidity language. Once the code of the contract is written, it can be compiled and deployed on the Ethereum blockchain. As mentioned before, in our case every action is to be stored in the blockchain. To do so, we use two kinds of Solidity smart contracts: `ShipmentManager` and `Shipment` contracts.

Shipment contract

A `Shipment` contract represents a single shipment in the simulation and stores every action performed on it. Thus, it contains the following elements:

- *id*: the shipment's identifier (i.e., the name of the shipment in the Anylogic simulation).
- *destination*: the X-Y coordinates of the final destination of the shipment.
- *actions*: the list of actions performed on the shipment.
- *shipmentManager*: a reference to its `ShipmentManager`, whose purpose will be explained later.

A `Shipment` also provides three callable methods:

- *addAction(action)* adds the specified action to the action list. It is only callable by the `ShipmentManager`.
- *getActionCount()* returns the current length of the action list.
- *getAction(index)* returns the action using its index in `actions`.

ShipmentManager contract

A `ShipmentManager` provides an interface responsible for managing `Shipment` contracts that characterizes shipments of a same application or context. It is composed of a single attribute and two methods:

- *shipmentsById*: a structure mapping shipment identifiers to their corresponding `Shipment` contracts. Therefore, it keeps a reference to every `Shipment` contract.

- *createShipment(shipmentId, destination)* creates a new `Shipment` by providing its *shipmentId* and its final *destination*, then appends it to the *shipmentsById* map. We assume that this function is called as soon as the shipment request is made in the simulation. Each creation produces a log event describing the new shipment.
- *addAction(shipmentId, action)* calls the *addAction* method of the `Shipment` contract corresponding to the provided *shipmentId* inside *shipmentsById*. Moreover, a log event describing the new action is emitted.
- *getShipment(shipmentId)* returns the address of the `Shipment` contract corresponding to the provided *shipmentId* inside *shimentsById*, or the null address (i.e., *0x0*) if it does not exist.

One could argue that there is no need for `ShipmentManager` contracts. Indeed, actions could be directly sent to corresponding `Shipment` contracts and the result would be the same. However, there are several justifications for this choice.

First, using only `Shipment` contracts implies that the agents must know each contract address, which means that some kind of database or file containing these addresses must be maintained and made available to all the agents, and that this system must be updated at each contract deployment. A `ShipmentManager` contract tackles this issue by keeping a reference to all the contracts and updating this list every time a `Shipment` is created: there is no need for an external system and the `ShipmentManager` address is all the agents of the simulation have to be aware of.

Secondly, having such an interface allows us to emit events on shipment creation and new actions for all shipments on a single contract. Otherwise, each `Shipment` contract would be responsible for emitting their events. If one wanted to monitor the actions realized on specific shipments, he would have to do it on a per shipment basis. In our solution, he would only have to monitor the events emitted by the `ShipmentManager` and filter out those he might not be interested in.

Finally, a `ShipmentManager` permits to perform verification on a per group of shipments basis. Indeed, the handling of some shipments might be constrained by specific rules (e.g., some actions on specific sets of shipments might be forbidden or allowed for particular agents). Therefore, access control mechanisms, data conformity enforcement, or any other kind of verification, might be directly implemented at the `ShipmentManager` level, thus allowing different granularity levels for controls.

3.3 Interactions Between Blockchain and AnyLogic

So far, our private blockchain network is up and running, and both `ShipmentManager` and `Shipment` contracts are written and deployed to the blockchain. However, since AnyLogic does not provide any framework to deal with Ethereum clients or smart contracts, external tools are needed in order to allow agents in the simulation to send actions to the blockchain.

3.3.1 The Web3j Library

Our first attempt was to use the web3j library [33], an open source Java library allowing the communication with Ethereum clients via inter-process communication (or IPC, usually requires that the Java application and the Ethereum client are running on the same computer), or remote procedure call (RPC) protocol (this option must be activated when starting the Ethereum client to authorize such interactions). With web3j, a Java program can easily connect to any available Ethereum client and conduct common transactions, such as sending Ethers—the Ethereum cryptocurrency—to another account, deploying a smart contract or calling its available methods.

One of the main interest of web3j is that it handles Java smart contracts generated from their current Solidity code, thus making it really straightforward to deploy smart contracts and call their methods. Let us take an example of a simple smart contract. This smart contract is called `MyStringStorageContract`, and provides an interface to store and modify a single string. Also, each time its string is modified, it should emit a log event. The Solidity code corresponding to such smart contract is reported in Fig. 4.

The constructor of the contract expects a string (lines 12–14) to initialize the `myString` variable, which will store the string persistently. Then, a method `getMyString` is provided to access the currently stored string (lines 16–19), as well as a `setMyString` method to modify it (21–24). The latter also emits an `OnStringChange` log event (defined on lines 7–10), which contains the old and new values of the string (line 24).

Web3j provides tools to generate a Java class used as a wrapper around the smart contract's interface, making it possible to interact with it using Java code [34]. An example of smart contract handling using web3j is illustrated by Fig. 5, with `MyStringStorageContract` as the Java class generated by the web3j tool from the Solidity code of Fig. 4, and imported into the Java project.

Here, the Java program (Fig. 5a) connects to an Ethereum client with RPC available at the address `http://localhost:8545/` (line 11). Then, the credentials (i.e., the agent's account address and public/private keys) are retrieved from a *wallet* file using a password (empty in our case) to decrypt the private key (line 14). A contract `MyStringStorageContract` is then deployed with the string `"Hello, you!"` as constructor parameter (lines 17–19). Note that we also have to specify the gas price and limit consumed by our transaction: in Ethereum, gas is used to evaluate the amount of computations needed by a transaction.[4] One client with access to

[4] In fact, users decide the maximum gas consumption of their transactions so that it can be rejected in case the limit is exceeded. Also, they suggest a price for each unit of gas (the lower, the better for them), but the choice to validate transactions with low gas price is left to the mining nodes' discretion, since it also affects their mining reward.

```
1  pragma solidity >=0.4.22 <0.6.0;
2
3  contract MyStringStorageContract {
4
5      string private myString;
6
7      event OnStringChange(
8          string oldString,
9          string newString
10     );
11
12     constructor(string memory _myString) public {
13         myString = _myString;
14     }
15
16     function getMyString() external view
17         returns (string memory _myString) {
18         _myString = myString;
19     }
20
21     function setMyString(string memory _myNewString) public {
22         string memory oldString = myString;
23         myString = _myNewString;
24         emit OnStringChange(oldString, myString);
25     }
26 }
```

Fig. 4 Solidity code of the smart contract `MyStringStorageContract`

the blockchain network could also load this smart contract by specifying its address, which can be retrieved once the contract has been deployed (line 23). Afterwards, it is possible to access the string contained in the smart contract with its `getMyString` method (line 26), and to replace it with the string `"Hello, world!"`, by calling its `setMyString` method (line 29).

The output of such program is shown in Fig. 5b. As expected, the address of the deployed smart contract is printed, and could be stored for later use, even by another client synchronized with the network, to manipulate the contract (line 1). Also, we can verify that the string retrieved directly after the contract deployment is indeed `"Hello, you!"` (line 2), and that it is effectively substituted for `"Hello, world!"` afterwards (line 3).

3.3.2 Action Managers and Architecture

Since AnyLogic simulations are developed in Java, the first idea was to import web3j in AnyLogic and deals with smart contracts directly inside the simulation. However, there is a conflict between the cryptographic library already present in AnyLogic and the one used by web3j to sign transactions, therefore making this option impossible. Thus, the alternative was to develop a program outside AnyLogic, which would be

```
1  // Each unit of gas is worth 10 Wei (1 Wei = 10^-18 ETH)
2  public static final BigInteger GAS_PRICE = BigInteger.valueOf(10);
3
4  // The cost of a transaction cannot exceed 6,000,000 gas
5  public static final BigInteger GAS_LIMIT = BigInteger.valueOf(6000000L);
6
7
8  public static void main(String[] args) throws Exception {
9
10     // Initiating connection to the Ethereum node
11     Web3j web3j = Web3j.build(new HttpService("http://localhost:8545"));
12
13     // Loading personal Ethereum wallet
14     Credentials wallet = WalletUtils.loadCredentials("", "wallet.json");
15
16     // Deploying the smart contract
17     MyStringStorageContract contract = MyStringStorageContract
18             .deploy(web3j, wallet, GAS_PRICE, GAS_LIMIT, "Hello,␣you!")
19             .send();
20
21     // Printing the contract's address
22     // It may be used to load the contract from the blockchain
23     System.out.println(contract.getContractAddress());
24
25     // Print the current content of the contract's string
26     System.out.println(contract.getMyString().send());
27
28     // Changing the string stored in the contract
29     contract.setMyString("Hello,␣world!").send();
30     System.out.println(contract.getMyString().send());
31 }
```

(a) Java code for MyStringStorageContract handling with web3j

```
1  0x4985f73ecaacd710a087388389d54522ad4af868
2  Hello, you!
3  Hello, world!
```

(b) Program's console output

Fig. 5 Example of smart contract handling using web3j and its corresponding console output

able to receive actions sent by the AnyLogic agents, then forward them as transactions to the Ethereum clients.

This intermediary program is called an *Action Manager*: it is actually an HTTP server aware of the agent's public/private keys, which is listening for POST requests containing action data from the simulation agent, then processes them into actual Ethereum transactions, and then send them to the Ethereum client via RPC. As illustrated in Fig. 6, this means that there are three different entities at the agent level: (1) the agent inside the AnyLogic simulation, (2) its Action Manager responsible for transaction forwarding, and (3) the ETH node which receives transaction from the Action Manager and communicates with the other ETH nodes of the blockchain network.

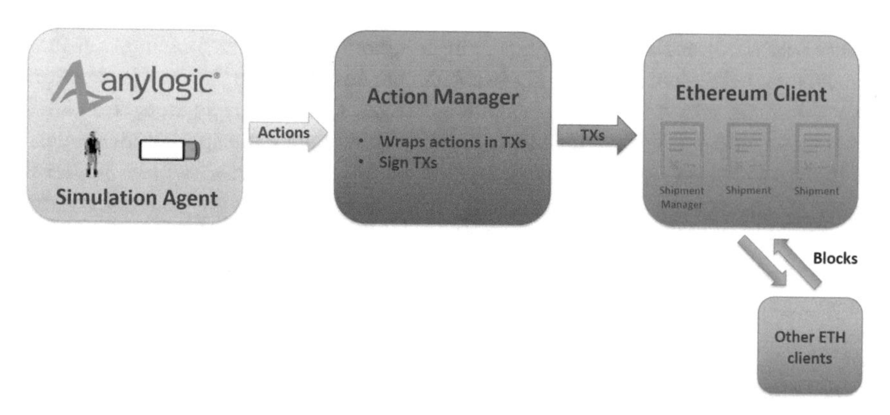

Fig. 6 Agent-level architecture (TXs = transactions, ETH = Ethereum)

3.3.3 Switching to `go-ethereum`

Although this solution is functional, it is not viable when dealing with a high number of agents, at least in our context of locally simulated network. As a matter of fact, a single Action Manager with `web3j` takes up to 200 MB of memory, and we have 40 of them running simultaneously (they are 40 agents—36 deliverers and 4 transporters—in the simulation). On a memory-constrained computer, this results in several `OutOfMemoryException`, causing the Action Managers to shut down.

In order to reduce their memory consumption, it was decided to switch from Java to Go programs. Indeed, the Go language offers very light HTTP requests handling, is the native development language of the Ethereum clients and thus benefits from the same features as web3j. Therefore, Action Managers were re-coded in Go and used the go-ethereum[5] library, the native official library for Ethereum clients. This change provided highly positive results, since running all the 40 Action Managers now only requires about 20 MB of RAM, against approximately 8 GB with our previous Java and web3j implementation.

4 Monitoring Lifecycle Properties for Hyperconnected Logistics

As we mentioned earlier, the hyperconnected logistics paradigm brings in many benefits. However, it also has drawbacks worth being stated. Because much more actors are involved in the transportation of a single parcel, this increases chances for unintentional errors or deliberate malicious actions. Indeed, the latter can be hidden more

[5]https://github.com/ethereum/go-ethereum.

easily since shipments change hands more frequently and because each one of them has its own individual *lifecycle*. By lifecycle, we designate the "correct" processing of a shipment, which is, in fact, best characterized by the steps a parcel should follow and the constraints that must be enforced throughout its transportation, similarly to business process management lifecycles [4, 35]. Moreover, it will be harder to know which actor is actually transporting a shipment at a particular moment, making it even more complex to efficiently track goods. As a matter of fact, introducing decentralization makes it more difficult to obtain accurate snapshots of the supply chain's global state. This brings up the need to state several "constraints" that must be enforced at all time for a single shipment or for the whole network, which are independent of its topology or the nature of the parcels being shipped.

On the other hand, with the blockchain setup described in Sect. 3, it becomes possible to outsource the whole task of keeping a consistent global record of actions to the blockchain infrastructure, and to only read the elements of the blockchain as they are appended, based on the events emitted by the smart contracts (see Sect. 3.2.2). This should be contrasted with other runtime verification techniques that also take into account the distributed nature of a system. A recent work by Falcone et al. tackles the tracing of a distributed component-based system [36]: the goal in this work is to produce a lattice of events that preserves the total ordering on each local component. On the contrary, our scenario requires a total ordering of events for each parcel, but these events typically take place at a different location every time they occur.

Previous work on decentralized monitoring is also worthy of mention [37–41]; however, we shall see that the type of properties we consider goes beyond the expressiveness of the finite-state automata, regular expressions and propositional LTL formulæ used in those works. Moreover, the structure of the hyperconnected supply chain is dynamic: hubs can join or leave the system at any time; this situation is harder to accommodate in classical decentralized monitoring, where the number of sites and connectivity between them is assumed to be known in advance.

Our proposed approach also differs from monitoring smart contracts [42], which can be instrumented and monitored like any other program. In the present work, we rather take the blockchain entries themselves as a source of events that is the subject of runtime verification, and this is done by a monitor that is external to the blockchain.

4.1 Lifecycle Constraints

In a system where the status of multiple parcels are tracked at the same time, it can be desirable to make sure that the events pertaining to a single instance of a parcel follow a correct sequence that correlates its various *pick-up* and *delivery* events. This gives rise to a first correctness property:

Property 1 A parcel must follow a precise lifecycle, where: 1. a *pick-up* event must be followed by a *delivery* event, and vice versa; 2. a *pick-up* must occur at the same location as the previous *delivery* (i.e. a parcel should not move in between); 3. the last event is a *delivery* at the expected destination. □

Deviations from this lifecycle can have multiple causes. For example, an employee may simply have omitted to scan a parcel when picking it up or delivering it. Other, more nefarious circumstances, such as a parcel ID being tampered or a parcel being moved to a new location without being scanned, can also be caught through violations of this lifecycle.

A parcel can also be *re-routed* while it is in transit. This means that a new destination is declared for this parcel; such a re-routing can be detected by witnessing a different set of x-y coordinates for the parcel's destination between two successive events. However, the number of re-routings for a given parcel should be limited by some predefined constant. This leads to the following constraint:

Property 2 The destination (i.e. target x-y location) of a parcel should change at most k times before the parcel reaches its declared destination. □

Additionally, one may want to avoid the destination of a parcel to oscillate back-and-forth between multiple destinations, leading to another constraint:

Property 3 Each re-routing must send the parcel to a new destination. □

Since the goal of a hyperconnected supply chain is to route parcels to their destinations efficiently, another property should stipulate that parcels are not expected to take detours in their way to their target.

Property 4 A parcel must always progress towards its destination. □

Monitoring such a property can be useful to detect parcels that start following a suspicious route that takes them away from their intended destination.

There exist other correctness properties which we only mention in passing due to lack of space. These include the following:

Property 5 A parcel cannot exceed some maximum speed over a single hop. □

There is indeed a reasonable upper bound to the distance a parcel can travel in a unit of time. Witnessing a parcel that exceeds this single-hop speed may indicate that its position has been spoofed, or than an error in the measurement of the position has occurred. A reverse reasoning could also be made on the minimum speed.

Property 6 Events about each parcel should be separated by no more than x time units. □

This property involves some form of "timeout"; it is used to issue a warning when no news about a parcel has been produced after some fixed amount of time. Lost parcels can be detected by monitoring this property.

The properties we have seen so far are relatively typical of the kind of constraints generally seen in Runtime Verification literature, and produce a Boolean verdict, i.e. the property is verified or not. However, the hyperconnected logistics scenario we introduced in this paper brings many other forms of computation over parcel events that fall outside of this definition. We now turn to more advanced properties that can also be monitored. First, we call "analytical" any property whose result is not a Boolean verdict. Such properties can be used, for example, to compute statistics on the state of the supply chain. For example:

Property 7 Compute the current number of parcels that are in transit, i.e. that have not reached their destination. □

Property 8 Compute the distribution of the time required to ship a parcel for a given source–destination pair. □

Property 9 Compute the number of parcels that transit through a given cell. □

All the correctness properties shown so far are considered "absolute", in the sense that the same condition must apply to all events, or all sub-streams for each parcel. For example, Property 2 imposes a maximum number of times a parcel can change its destination; this number is known in advance and applies to all parcels. There exist other properties where what is considered correct actually depends on aggregations calculated on windows of past events of the same stream. That is, a static reference cannot be established in advance. We call these properties "self-correlated", in the sense that at least some of their parameters are based on the content of the stream itself.

A prime example of such a property relates to the shipping duration of parcels across a single hop. It is hard to determine in advance what constitutes a "normal" shipping time for each hop in a hyperconnected grid; moreover, this normal duration may drift over time, making a static reference value eventually deprecated. However, one may be interested in large deviations in shipping time over what was customarily seen in the past; this can be expressed by a property such as this one:

Property 10 The average shipping time of the past m parcels on a given hop must not exceed by a factor k the average time for the previous n parcels on that same hop. □

4.2 BeepBeep: A Complex Event Processing Library

Now that we have specified relevant properties for hyperconnected logistics, we can use the data stored in the blockchain as an event source stream to verify them. First, however, we need to present BeepBeep, the complex event processing library used to implement the previously mentioned properties in this chapter.

4.2.1 Complex Event Processing

Event streams have become an important part of the mass of data produced by computing systems. They can be generated by a myriad of sources such as sensors [43–45], business process logs [46], instrumented software [47–49], financial transactions [50], healthcare systems [51], network packet captures [52], and, as in the present chapter, blockchain transactions. The ability to collect and process these event streams can be put to good use in fields as diverse as software testing, data mining, and compliance auditing.

Complex Event Processing (CEP) can loosely be defined as the task of analyzing and aggregating data produced by event-driven information systems [53]. A key feature of CEP is the possibility to correlate events from multiple sources, occurring at multiple moments in time. Information extracted from these events can be processed, and lead to the creation of new, "complex" events made of that computed data. This stream of complex events can itself be used as the source of another process, and be aggregated and correlated with other events.

Event processing distinguishes between two modes of operation. In online (or "streaming") mode, input events are consumed by the system as they are produced, and output events are progressively computed and made available. It is generally assumed that the output stream is monotonic: once an output event is produced, it cannot be "taken back" at a later time. In contrast, in offline (or "batch") mode, the contents of the input streams are completely known in advance (for example, by being stored on disk or in a database). Whether a system operates online or offline sometimes matters: for example, offline computation may take advantage of the fact that events from the input streams may be indexed, rewound or fast-forwarded on demand. However, in our blockchain-based case, whether the processing occurs in online or offline mode does not matter since our method to monitor transactions makes no distinction between "live" and "stored" blockchains.

4.2.2 An Overview of BeepBeep

BeepBeep 3 is an event stream processing engine implemented as an open source Java library.[6] It is organized around the concept of *processors*. In a nutshell, a processor is a basic unit of computation that receives one or more event traces as its input, and produces one or more event traces as its output. BeepBeep's core library provides a handful of generic processor objects performing basic tasks over traces, summarized in Fig. 7; they can be represented graphically as boxes with input/output "pipes".

The main processors provided by the core library are described as follows. The `ApplyFunction` processor is such that, given a function f, it simply applies f to each input event e and returns $f(e)$. A function can have multiple input arguments; in this case, the corresponding processor has as many input pipes as the input arity of the function. The `Cumulate` processor calculates a cumulative "aggregation"

[6]https://liflab.github.io/beepbeep-3.

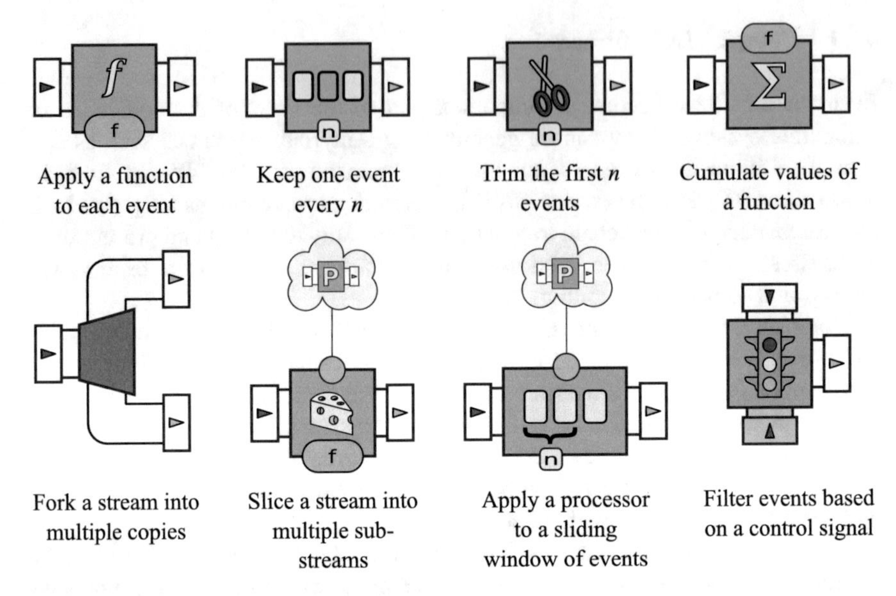

Fig. 7 BeepBeep's basic processors

on the events received so far. A user-definable function f is applied to the stream; this function must have two input arguments. If we note the previous value returned by the processor as a, upon the arrival of a new event b, the event output by the processor will be $f(a, b)$. An initial value a' must be defined for producing the first event. The role of the `Trim` processor is to delete a fixed number of events from the beginning of a stream. The number of events to delete is specified as an argument when the processor is created. Conversely, the `Pad` processor inserts an event a predefined number of times, before starting to output the events produced by some other processor P. The `Decimate` processor rejects events from an input stream at periodic intervals. This task can be performed in two ways: on the basis of a fixed number of events (`CountDecimate`), or based on a fixed time interval (`TimeDecimate`).

The purpose of the `Fork` processor is to divide the original stream into several identical copies. This division makes it possible to perform several separate calculations on the same stream. The `Filter` processor allows the user to keep or delete events from an input stream arbitrarily. In its simplest form, a `Filter` has two input *pipes* and an output *pipe*. The first input channel receives the stream of events to be filtered and the second channel receives Booleans. If the value of the Boolean at the position n is $\top$ (true), the event at the position n in the input stream is sent to the output.

As its name implies, the `Window` processor performs a calculation on a sliding window of events of the input stream. A window of n successive events is accumulated, and then fed to some other processor P. The output of P on this window is returned as the output event associated to this window. The arrival of a new event pushes the oldest one out of the window, and the computation restarts for this new window. An interesting feature of BeepBeep is that P is not restricted to classical aggregations such as sums and averages; any processor can be used as the window computation, including non-numerical processes. Finally, the `Slice` processor is one of the most complex of the core library. It uses a function f to separate an input stream into several sub-streams. Each of these sub-streams is sent to a different instance of some processor P, and the output of each copy is aggregated into an associative map. As a simple example, one could use a slice processor to separate a business process log into separate instances of the process (based on some process ID), and perform some computation P on each instance separately.

In order to create custom computations over event traces, BeepBeep allows processors to be *composed*; this means that the output of a processor can be redirected to the input of another, creating complex processor chains. Events can either be *pushed* through the inputs of a chain, or *pulled* from the outputs, and BeepBeep takes care of managing implicit input and output event queues for each processor. In addition, users also have the freedom of creating their own custom processors and functions, by extending the `Processor` and `Function` objects, respectively. Extensions of BeepBeep with predefined custom objects are called *palettes*; there exist palettes for various purposes, such as signal processing, XML manipulation, plotting, and finite-state machines. Over the past few years, BeepBeep has been involved in a variety of case studies [43, 54–58], and provides built-in support for writing domain-specific languages [59]. A complete description of BeepBeep is out of the scope of this paper; the reader is referred to a recent tutorial [60] or to a recent textbook about the system [14].

4.3 Connecting Ethereum to BeepBeep

In order to enable the verification of properties presented previously, BeepBeep must have access to the information contained in the Ethereum blockchain, and BeepBeep's core does not provide this feature.

As mentioned earlier, web3j allows, among other things, the deployment of new smart contracts on a blockchain, to send transactions towards them, and to listen to *log events*, a special kind of events emitted by smart contracts which are written in the log of the Ethereum blockchain. By doing so, it is possible to use web3j to catch any event emitted by a specific contract and use it for further processing.

Web3j and BeepBeep being both written in Java, a first step in was to implement an Ethereum palette for BeepBeep.[7] The palette is actually implemented as a BeepBeep wrapper over functionalities provided by web3j, and makes it possible to use log events emitted by a blockchain smart contract as inputs for a BeepBeep chain of processors. The following code fragment shows how a special BeepBeep processor,

[7] https://github.com/liflab/bb-palette-blockchain.

called CatchEthContractLogs, can be instructed to connect to a blockchain and listen to the events that are appended to it:

```
CatchEthContractLogs source = new CatchEthContractLogs(
  "http://localhost:8545", "0x6702413C52c8Cf0fc5f", true);
```

The first parameter is the actual URL of the RPC-enabled Ethereum node; in our example, the node is located on the host machine and is accessible through a specific TCP port. The second parameter is the address of the smart contract one wishes to listen to. Finally, the Boolean parameter true indicates that the processor will catch every event in the chain from the earliest block. Setting it to false would result in the processor only reporting *new* elements added to the blockchain after the processor is created.

This processor acts as a BeepBeep Source processor, and in fact is a descendant of that class. From this point on, instances of CatchEthContractLogs can be used like any other BeepBeep processor; for instance, they can be connected to downstream processors that perform some form of processing on the blockchain events. It is designed in such a way that upon calling its start method[8] all entries already in the blockchain will be pushed downstream one by one in a single burst (or not, if the third parameter of the constructor is set to false).

Since a blockchain can contain entries of various types, each having its own data structure, one may be interested in filtering this blockchain and keep only entries of a certain type. To this end, the palette also defines a BeepBeep Function object called GetEventParameters. This function is instantiated with what is called a *filtering object*; any element of the blockchain that matches the name and structure of that object will be kept; others will be filtered out. For example, let us assume that an event called Instructor is emitted on the blockchain, and contains a name string parameter and an age integer parameter. Such event would be declared in a Solidity smart contracts as follows:

```
event Instructor(
   string name,
   uint age
);
```

Using BeepBeep and its Ethereum palette, it is possible to retain only events of this type using the following code block:

```
ApplyFunction filter = new ApplyFunction(new GetEventParameters(
  new Event("Instructor",
  Arrays.asList(new TypeReference<Utf8String>() {},
    new TypeReference<Uint256>() {} ))));
```

This piece of code instantiates an ApplyFunction processor, whose function is an instance of GetEventParameters. This function itself is given an instance of an Event object (a class provided by the web3j library) that defines the name and internal structure of the entries to look for in the blockchain.

[8]A method provided by BeepBeep's top-level Processor class, and which can be overridden to perform various startup tasks, depending on the specific processor. The default behavior is to do nothing.

4.4 Implementing Properties with BeepBeep

It is now possible to implement the properties presented earlier and summarized in Table 1 as chains of BeepBeep processors. Most of them will be illustrated by the corresponding processor chain. Following BeepBeep's graphical conventions, specific colors will refer to events of a specific type: tuples (yellow), character strings (purple), numbers (teal), Boolean values (grey-blue), associative maps (dark blue) and composite data structures such as sets and lists (pink with polka dots).

Property 1 is best modeled by BeepBeep's `MooreMachine` processor, which allows users to define instances of Moore machines [61], where each state is associated with an output symbol; this is shown in Fig. 8. The Moore machine itself is illustrated as the contents of box #1; as one can see, the states of this particular machine are labeled with values $\top$ (true), $\bot$ (false) and "?" (inconclusive). Transitions in this machine are associated to functions to be evaluated on an incoming event; from a given state, a transition fires if the corresponding function evaluates to true on the given event.

A Moore machine also has *state variables*, which can be modified when taking a transition. For example, the transition associated to box #2, if taken, will associate to an internal variable called x the value of attribute *loc_x* in the current event (and similarly for y). These state variables can also be queried in a transition, such as in box #3, which checks that the values of *loc_x* and *loc_y* in the current event are equal to the values saved in state variables x and y.

Since this lifecycle applies to each parcel individually, a `Slice` processor creates one sub-stream for each parcel, by creating one instance of the machine for each

Table 1 A summary of the lifecycle properties from Sect. 4.1

Name	Description
Property 1	A *pick-up* event must be followed by a *delivery* event, and vice versa
	A *pick-up* must occur at the same location as the previous *delivery*
	The last event is a *delivery* at the expected destination
Property 2	Destination of a parcel should change at most k times before the parcel reaches its declared destination
Property 3	Each re-routing must send the parcel to a new destination
Property 4	A parcel must always progress towards its destination
Property 5	A parcel cannot exceed some maximum speed over a single hop
Property 6	Events about each parcel should be separated by no more than x time units
Property 7	Compute the current number of parcels that are in transit
Property 8	Compute the distribution of the time required to ship a parcel for a given source–destination pair
Property 9	Compute the number of parcels that transit through a given cell
Property 10	The average shipping time of the past m parcels on a given hop must not exceed by a factor k the average time for the previous n parcels on that same hop

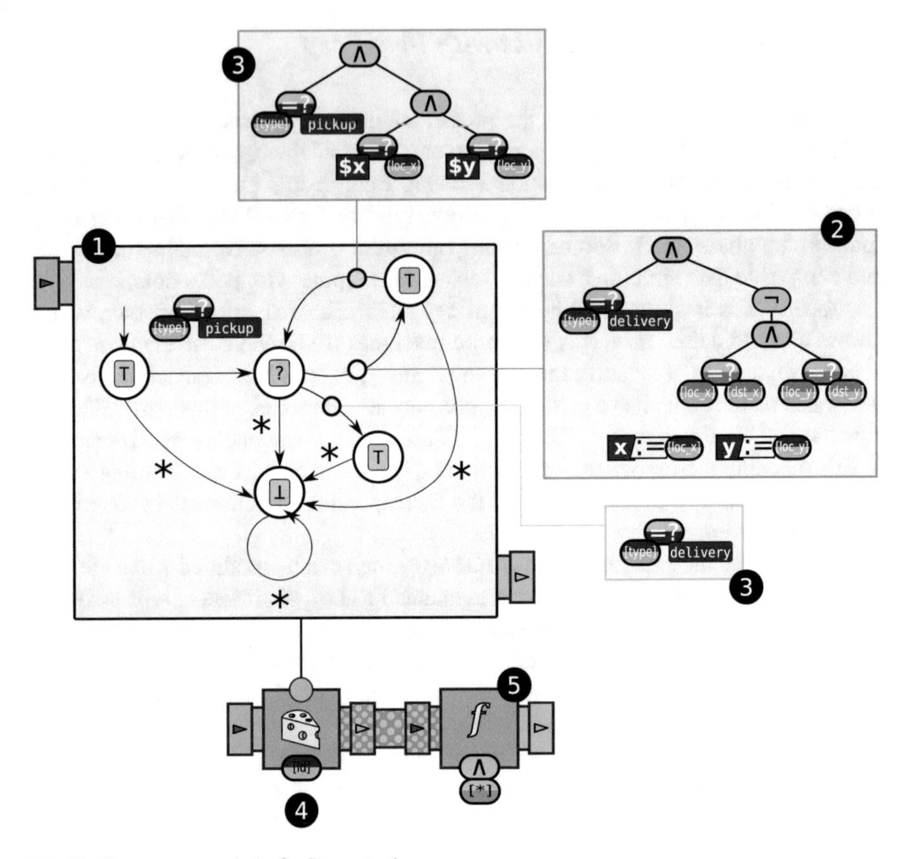

Fig. 8 The processor chain for Property 1

distinct parcel *id* (box #4). The output of the slice processor is an associative map between sub-stream *keys* (in this case, parcel IDs) and the last event output by each sub-slice processor (in this case, the Boolean value produced by the corresponding Moore machine). To turn this chain into a correctness property, the last step is to extract the values out of the associative map, and to perform the conjunction of all these values; this is the task done by box #5. The end result is a stream of Boolean values that will return false ($\perp$) exactly when one of the parcels has violated its lifecycle.

Property 2 enforcement can be taken care of by the chain of Fig. 9. From a given event, the *destination_x* and *destination_y* attributes are extracted to create an x-y tuple (#1 in the figure). These tuples are accumulated into a set (box #2), the cardinality of this set is extracted (#3) and compared to a predefined value k (#4). Such a process is repeated for every parcel *id* through a `Slice` processor (box #5), and the conjunction of the values of the resulting associative map is taken (#6).

Property 3 can be enforced by the processor chain shown in Fig. 10. The part of the chain at #1 extracts the x-y destination of an event; chain #3 at the top detects

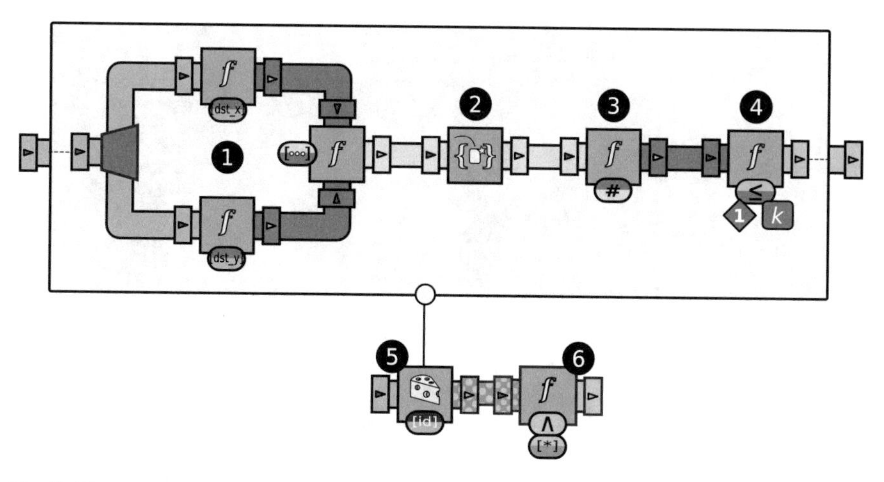

Fig. 9 The processor chain for Property 2

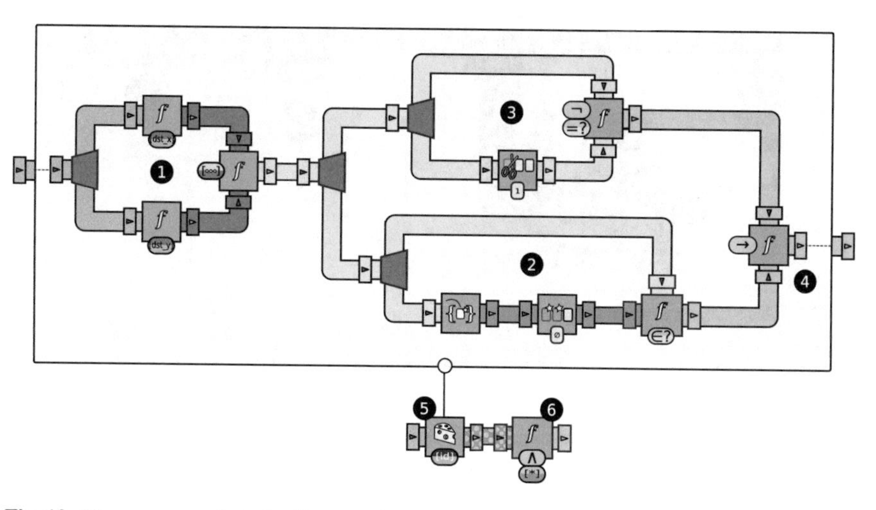

Fig. 10 The processor chain for Property 3

when two successive events have a different destination. Chain #2 at the bottom accumulates the destination tuples into a set, and checks if the current destination was already present in the *previous* set; this is done by inserting a single instance of ∅ into the stream, in order to offset the element and the set to be compared by one event. The Boolean streams in both branches are combined into an implication (#4); the result is a stream that returns ⊥ if and only if a change of destination occurred and the new destination was already present in the set of past destinations. Boxes #5 and #6 split a stream for each parcel *id* and take the conjunction of all sub-streams, as we have seen before.

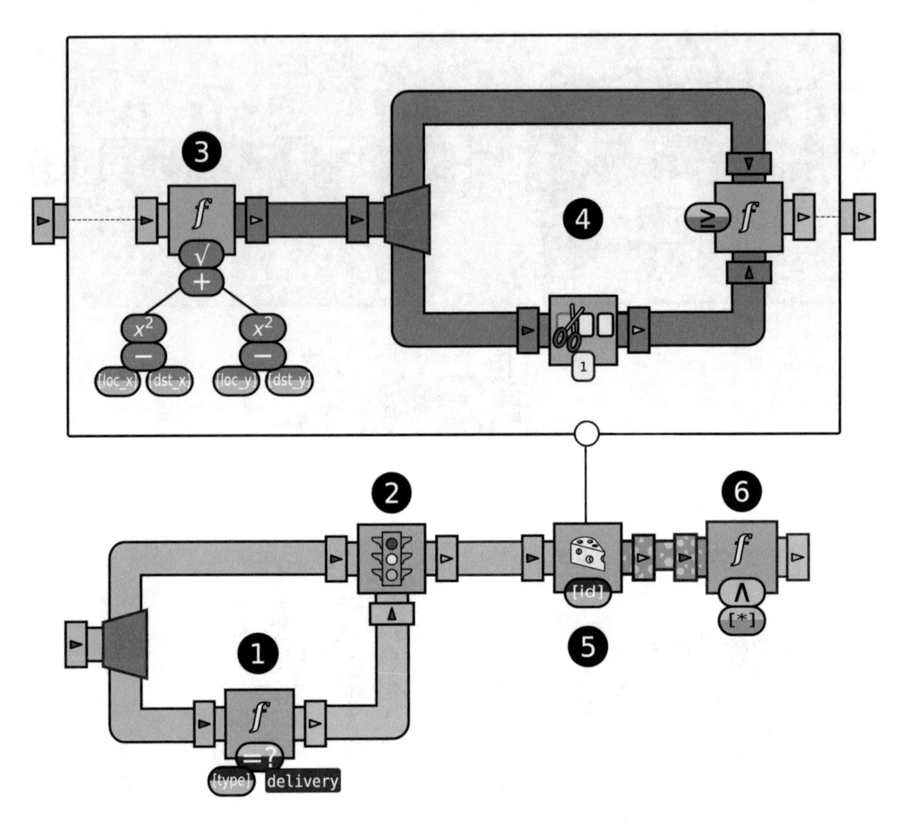

Fig. 11 The processor chain for Property 4

Property 4's chain of processors implementation is shown in Fig. 11. First, any event that is not a *delivery* is filtered out of the stream; this is done by processors #1 and #2. The remaining stream is sliced for each parcel *id* (#5). For each, the current distance to its destination is computed (#3), and the distances between two successive events are compared (#4). Therefore, processor #5 outputs an associative map where keys are parcel *ids*, and values are Booleans indicating whether each parcel has progressed towards its destination in the last two events concerning it. Box #6 takes the conjunction of all these values to turn the chain into a correctness property.

Property 7 can be calculated by using a modified version of Fig. 8 (not shown). Instead of producing Boolean symbols, the Moore machine of box #1 can be adapted to output the symbol 1 as long as a parcel is in transit, and 0 when it reaches its destination. Instead of the conjunction, box #5 can compute the sum of these values, which will correspond to the number of "live" parcels that are still in transit. A more complex analytical property is the following:

For Property 8, we consider a simplified version of the property where we assume that a parcel is not rerouted. The difficulty here lies in the fact that the source–destination pair is seen only in the first event for a given parcel; subsequent events

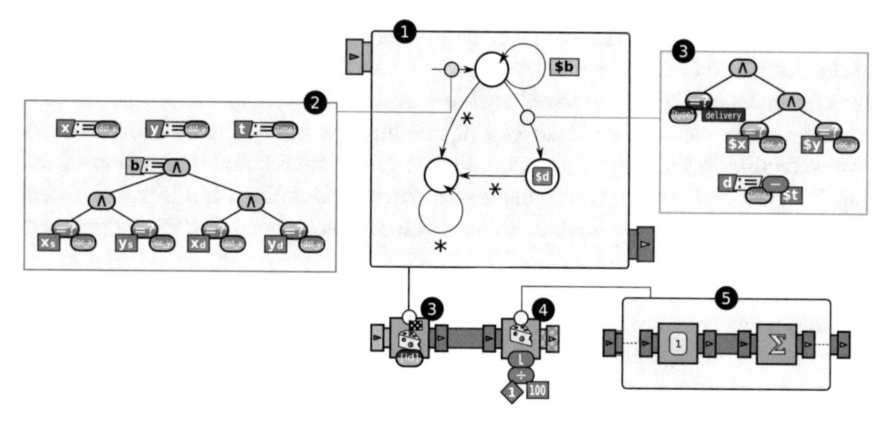

Fig. 12 The chain of processors for Property 8

are linked to it by the parcel ID, but will have a different pair of x-y coordinates for their source.

This can be calculated by the processor chain in Fig. 12. Due to lack of space, we only briefly explain its operation. A Moore machine is instantiated for each parcel; each instance of the machine will output the shipping duration of a parcel if this parcel had the intended source–destination pair, and nothing otherwise. The slice processor in box #3 creates one instance of this machine for each distinct parcel ID; the flag pictogram designates the variant of `Slice` that outputs the events produced by each internal processor directly, without accumulating them into an associative map. This produces a stream of shipping durations, one for each parcel having the desired source–destination pair. These numbers are accumulated into yet another `Slice` processor (box #4); this time, the slicing function takes a duration d and creates one slice for each value of $\lfloor \frac{x}{100} \rfloor$. In other words, a first slice will gather the durations between 0 and 100, another will gather the durations between 100 and 200, and so on. The processor that runs on each slice is a simple counter (box #5). As one can see, the output of this chain is a stream of associative maps; in each map, an entry $d \mapsto n$ indicates that there have been n parcels so far whose shipping time belongs to category d. This map could then be used for further processing, for example by displaying it in the form of a histogram.[9] One could also imagine variants of this property, such as computing the distribution of the shipping time for all segments corresponding to a single hop in the grid.

Property 10 can be verified in two steps. The first is to determine the shipping duration of each parcel on each single hop. This is done by defining a Moore machine, as shown in the left part of Fig. 10. Upon creation, this machine saves into state variables x, y and t the x-y coordinates and timestamp of the event. Assuming that Property 1 is enforced, this first event is a *pick-up*. For the same reason, the next

[9]Using BeepBeep's Mtnp palette, this would amount to placing a single other processor box at the end of the current chain.

event is a *delivery*; upon taking this transition, the machine saves the x-y coordinates of the destination in state variables x' and y', and the duration (difference between current timestamp and t) in variable d. It then emits a symbol, which in this case is a *tuple*. The value of attribute *hop* of this tuple is itself a list of four numbers, corresponding to the pair of x-y coordinates of the source and destination of this hop. The value of attribute *dur* contains the shipping duration for this parcel on this particular hop. The end result is a stream of tuples, each containing the endpoints of a hop and a shipping duration.

The next step is to compare, for a given hop, the average of the durations of the n previous packets to the average of the m that precede them on the same hop. A chain of processors doing such a computation is shown in the right half of Fig. 13.

First, the input stream is sliced based on the value of the *hop* attribute (box #7); in each slice, the value of *dur* is extracted (box #1). Then, the stream of events is sent into a *sliding window* computation (box #2). Such a computation requires two parameters: the width of the window (labeled n), and a processor to be applied on each window, represented by box #5. In this particular case, the computation to be performed is the running average over all events of the window. A similar computation is done on a second copy of the stream for a window of width m (box #4); because of the presence of the `Trim` processor (box #3), these two windows are offset by n events. The net effect is that processors #2 and #4 calculate the running average over two windows: one from the "present" (W_1), and another from the "past" (W_2).

One can see the average over these two windows as "trends", $\overline{w}_1$ and $\overline{w}_2$. The last step is to compare the running average over these two windows, and to ensure that they do not deviate too much. This is the purpose of processor #6, which computes the ratio $\overline{w}_1/\overline{w}_2$, and ensures that it does not exceed some predefined value k. The net result is a stream of Boolean values; value $\perp$ is emitted if and only if, for a given hop, the average shipping time of the last m parcels exceeds by a factor k the average shipping time of the n parcels before them. Processor #8 computes the conjunction of these values for all hops.

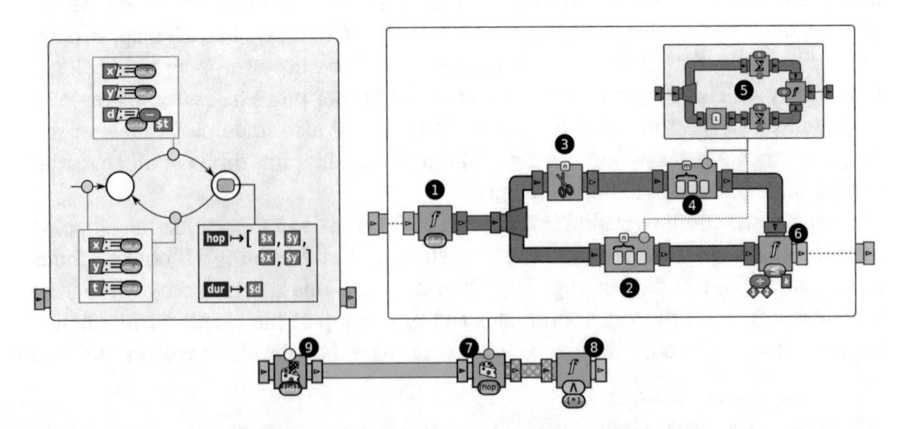

Fig. 13 The chain of processors for Property 10

This pattern differs from others seen before in a few respects. First, an error is not caused by a single event, but rather by the manifestation of a deviation of a trend calculated over multiple events. Second, what is considered "incorrect" is relative to a window of past events computed on the *same* stream. Finally, note that each hop is considered in isolation: the average shipping time will generally differ from one hop to another, so each hop is compared against its own past.

This pattern is actually a generic one. By replacing the processor in box #5 by another, and by changing the distance function in box #6 accordingly, it is possible to compute various trends over a stream of events, and to detect deviations of these trends in real time. These trends do not even need to be numerical scalars, provided that a suitable distance metric is given to compare them. As a matter of fact, Fig. 13 is an instance of the *self-correlated trend distance* workflow pattern, a concept that has been introduced in a previous study of data mining on event streams [62].

Specifying chains of processors with BeepBeep

To illustrate BeepBeep's coding aspects, Fig. 14 contains the code of the `main` method of a Java program monitoring Property 1. Here, information about each shipment's steps are stored in a CSV file. Therefore, the source of events is a `ReadLines` processor that outputs an event for each line of the CSV file (lines 3–5). A `TupleFeeder` processor is then connected to the `reader`, converting CSV lines to tuples, i.e., sets of CSV field name-value pairs (line 6–7). These tuples are then forwarded to a `Pump` (lines 9–10), which forces the input source to *consume* its events (i.e., each line of the CSV file is read and consumed by the chain of processors until the end of the file). Afterwards, the pump is connected to the processor `ParcelLifecycle` (lines 12–13), which actually depicts the behaviour of the processor chain in Fig. 8. For the sake of simplicity, the content of this processor is not detailed here, but can be found in the corresponding Java project,[10] as well as the code of the other properties of this chapter. A `Printer` processor is then connected to the chain, printing to the console the events it is fed with (lines 15–16). Finally, the `pump` is started to begin the processing of the chain (line 18).

5 Results and Discussion

So far, we have described our blockchain setup and how we could monitor properties on events produced by smart contracts' log events. In this section, we discuss the results of such implementations, in terms of provided features, limits, and performance issues.

[10] https://github.com/sylvainhalle/supplychain.

```
1  public static void main(String[] args)
2  {
3          String filename = "data/example_lifecycle_1.csv";
4          ReadLines reader = new ReadLines(
5                  ParcelLifecycleExample.class.getResourceAsStream(filename));
6          TupleFeeder tuples = new TupleFeeder();
7          connect(reader, tuples);
8
9          Pump pump = new Pump();
10         connect(tuples, pump);
11
12         ParcelLifecycle pl = new ParcelLifecycle();
13         connect(pump, pl);
14
15         Print print = new Print().setSeparator("\n");
16         connect(pl, print);
17
18         pump.run();
19 }
```

Fig. 14 Example of processors' chaining with BeepBeep

5.1 Applying Blockchain Technology to Supply Chain

In order to assess our blockchain-enabled simulation, we need to sum up the benefits of the blockchain application to our supply chain scenario, list matters that are still not dealt with in this solution, and discuss performance issues, in terms of blockchain speed and size.

5.1.1 Enforcing a Shared Shipment Tracking System

The simulation described earlier allows various agents to share the same blockchain network in order to store information on shipments they operate. This has several benefits.

The first is the uniqueness of parcels' virtual identities and data availability. Most of the time, companies involved in supply chain processes do not share the same information system. This results in information replication and/or information silos, i.e., tracking data are separated across the parties and one cannot access others' information directly. With blockchain technology, it is possible to provide a shared and trustworthy information system (which is not managed by a single entity) so that data is available to everyone at each step, therefore improving transparency for businesses as well as consumers.

Secondly, such system is completely agnostic as to the agents' nature. This means that it is very easy to add new agents from new companies to participate in the shipping. All it takes is the creation of an account for each new agent, since agents' transactions are dealt with in the same way, whatever their company affiliation or their nature (deliverer or transporter in the case of our simulation). In fact, agents need not even to be human beings, they could be IoT devices operating shipments automatically inside a hub. All they require is an Ethereum account. Clearly, if a company or a parcel requires specific verification steps, the system could be adapted so that additional smart contracts might be deployed to enforce the required controls.

5.1.2 Blockchain Size and Data Replication

Illustratively, we ran the simulation for 100 seconds, which corresponds to about 4 hours in the simulated city, and we monitored the evolution of the blockchain size. Figure 15 shows the evolution of the blockchain size as a function of time 15a, and as a function of the number of actions 15b. After 100 seconds, the blockchain size reaches 532 KB, which corresponds to 318 actions sent by the AnyLogic agents. Since the blockchain size evolves linearly against the action number, we can deduce that a transaction containing an action takes 1.6 KB in average.

For large shipping companies like Amazon or UPS, this could mean that several thousands of gigabytes of data are potentially added every single day. Although it might not be an issue using current storage technologies, it is important to mention that every full node in a blockchain network stores the same blockchain, and that the security of blockchain networks relies on the fact that every full node must have a copy of it. Since their number is expected to be quite high (e.g., in the Bitcoin network, there are currently about 9000 full nodes [63], and about 8300 for the Ethereum main net [64]), this means that there might be a huge amount of redundant information stored.

This problem might be tackled by storing the actual data in a regular distributed database instead of directly in the blockchain. To ensure the integrity and security of the data, the hashes of each action might be stored in the blockchain. This means that transactions would have a fixed size (due to the hash). Additionally, if the data volume ever became too large, one could decide to reinitialize the blockchain without losing the actual data.

However, this implies that an entire external system must be set up and managed (in addition to the blockchain network) and most likely by a single entity, which might seem contradictory since the blockchain principle is that there is no need for central authorities or intermediaries. To avoid this last issue, note that it might be possible to implement this external database system in a P2P way, where each blockchain node would also be a database node, and where data is spread all over the nodes, with a minimal but necessary replication for security, availability and resilience purposes, like in BigchainDB [31].

5.1.3 Access Control and Confidentiality

In our simulation, any agent connected to the network is able to send transactions and therefore provide tracking information about shipments. Each agent also has access to all the actions that have been or will be performed on the shipment. Although

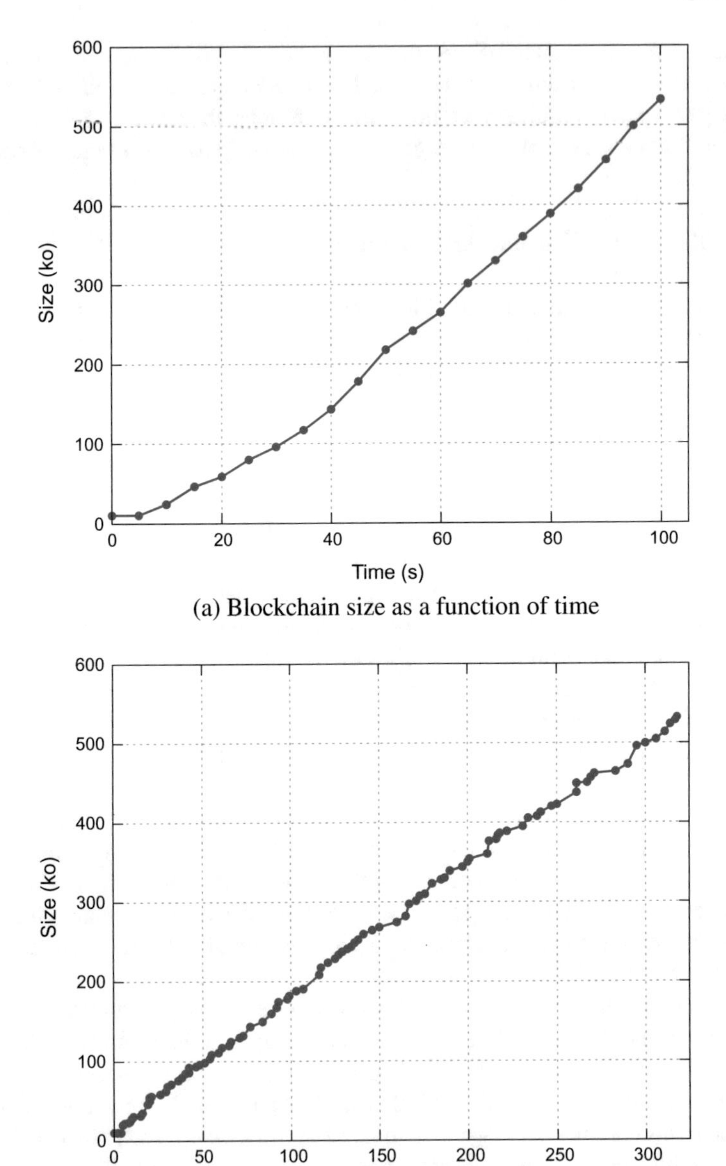

(a) Blockchain size as a function of time

(b) Blockchain size as a function of the number of actions

Fig. 15 Blockchain size analysis

it is actually one of the main benefits of blockchain technology, this might also be quite dangerous. Any malicious agent could potentially send fake information to the network for various reasons (e.g., theft, corrupting sensible data, overloading the network), and could also access information he should not.

This issue can be overcome by deploying smart contracts applying various access controls [65], depending on the agent's address (which is directly linked to its public key, so it is not possible to fake it, as it would also fake the transaction signature). These smart contracts could, for example, allow only an updatable list of account addresses to append actions. However, although this solution protects against malicious write attempts, it does not resolve the issue of potential forbidden read access. Here, only two options appear possible currently: (1) encrypt sensible data and distribute the keys to the agents, (2) store data in an external system and only save data hashes in the blockchain (as we mentioned in the previous section).

5.1.4 Performance

One of the main problem of the Ethereum network is that it supports a very low transaction rate (expressed in TPS, i.e., transactions per second): a few dozen TPS at most on the main net. In this simulation, approximately 3.2 transactions are processed per second. Although Ethereum network's speed would not be troublesome in this case, this could be a major issue in real full-scaled supply chain scenarios, since thousands of transactions might occur at the same time across the world. A common practice among cryptocurrency communities is to compare their transaction rate to Visa, which claims that its "software runs on powerful servers, mainframes and data storage systems that are capable of processing more than 65,000 transaction messages per second" [66], and manages to handle an average rate of approximately 4000 TPS, calculated from their 124.3 billion transactions processed over the year 2018 [66], which is way faster than Ethereum.

However, other blockchain networks might be used to solve this problem. For example, EOS.IO is a blockchain platform based on "a blockchain architecture that may ultimately scale to millions of transactions per second" [67]. In practice, EOS recently reached a peak of nearly 4000 TPS [68], which is still far from Visa's advertised peak capabilities, but almost enough to handle real supply chain scenarios (more details on this are given in Sect. 5.2). Moreover, Futurepia, a blockchain network under development specialized in social media platform solutions for decentralized applications (also called DApps, i.e., applications deployed over a blockchain network), enables a speed of 300,000 TPS [69], which is more than enough for our kind of scenario.

As a matter of fact, there are huge differences between major blockchain networks' transaction rates and those mentioned above. This is because most of them use the Proof-of-Work consensus algorithm, a very expensive mechanism in terms of computation and time. EOS and Futurepia tackle these limitations by adopting respectively Delegated Proof-of-Stake (DPoS) [70] and Dual Delegated Proof-of-Stake (DDPoS) [71] consensus, which are much faster and more energy-saving.

5.2 Property Verification Using BeepBeep

A second phase of experimentation is concerned with measuring the performance of the monitors that watch the lifecycle of each parcel. Indeed, the whole point of the blockchain approach is not only to store all events into a distributed ledger, but to use this ledger to make sure that each parcel follows its intended route.

Therefore, to assess the scalability of the proposed approach for property monitoring, we ran each of the processor chains illustrated in the paper on simulated sequences of 500,000 blockchain transactions, by varying the number of parcels simultaneously in transit between 10 and 10,000 and the number of hops taken by each parcel between 10 and 100. Each parcel was given a probability of $\frac{1}{5}$ of being rerouted upon taking each hop of its path. The properties were monitored on a computer with an Intel CORE i5-7200U 2.5 GHz running Ubuntu 18.04, with 1847 MB allocated to the Java virtual machine. All experiments and data are available for download [72] as a self-contained instance of the LabPal experimental environment [73]. In total, 56 experiments were run, generating 168 individual data points.

The first element to measure is how well the monitoring approach scales with the number of events received. An example is shown in Fig. 16, for the Decreasing distance property. As one can see, the cumulative execution time follows a precise linear trend; this means that the time required to process a single blockchain event is independent of the number of events received in the past. This condition is essential for the lifecycle monitoring approach to be usable in practice. Although not shown, all other properties exhibit a similar behavior with respect to the total number of events.

Another element worthy of consideration is how each property scales with respect to the number of hops (distinct events in the lifecycle of each parcel), and the number of parcels occurring simultaneously in the blockchain. Figure 17 shows the throughput, in number of input events consumed per second, for each of the properties and a varying number of hops per parcel. Figure 18 shows the throughput for each of the properties, this time by varying the number of live parcels in the simulation.

Although the results display some variability, one can observe that few properties seem to be negatively impacted by an increase in the number of parcels or in the length of each parcel's lifecycle. This can be explained by the fact that most of these properties perform a computation step upon every event of the blockchain; apart from the number of distinct slices that can be created, whether an event belongs to a parcel or another does not incur a different cost. However, more parcels usually means more events per unit of time; for example, a simulation with 100 parcels will probably generate events ten times faster than a simulation with 10.

Therefore, a more interesting measurement is to divide the global throughput by the number of parcels in the simulation; this value gives an indication of the number of events per second that each parcel can generate; doing so on the results of Fig. 17 reveals that this number does decrease with the number of parcels. As a matter of fact, the slowest of all properties can handle 14967.849 events per second for 10,000

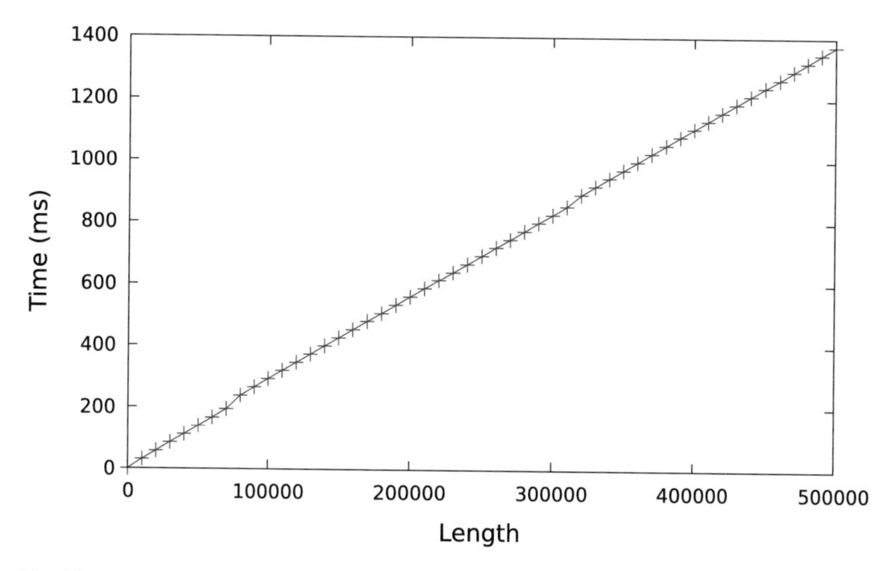

Fig. 16 Cumulative processing time with respect to number of events, for the Decreasing distance property, with 10,000 parcels and 100 hops

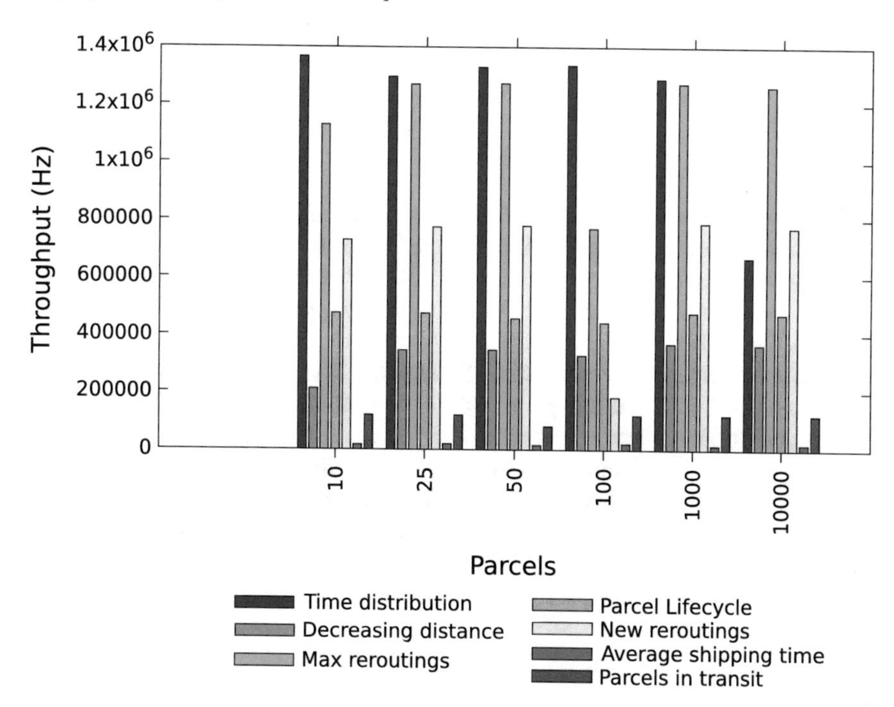

Fig. 17 Throughput for each property, in thousands of events per second, by number of parcels, for 100 hops

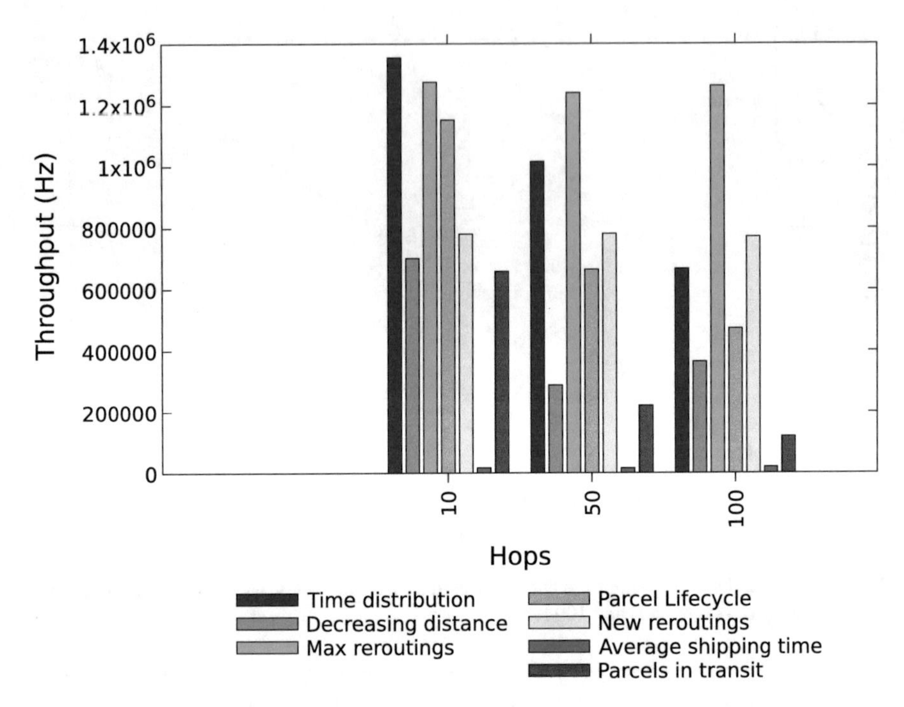

Fig. 18 Throughput for each property, in thousands of events per second, by number of hops, for 10,000 parcels

parcels. This means that each individual parcel could generate about 1.5 updates per second into the blockchain and still be handled in realtime.

These figures should be put into context by comparing them to estimates from an ongoing project in which one of the authors of this chapter (Montreuil) is involved in Shenzen, China. In this hyperconnected infrastructure, it is expected that in a normal day, one million parcels would transit through the system, each following a route containing between 2 and 8 hops, and generating an estimated total of roughly 80 million events. Peak days are estimated to five times this volume, which yields a worst-case figure of 400 million events per day. Handling these events in real time requires a throughput of 4600 Hz. Figures 17 and 18 show that the processing power available on a single computer largely exceeds this value; even the slowest monitor (Decreasing distance) has a throughput of over 13,000 Hz, enough to process close to three times the estimated peak of activity.

6 Conclusion

The Physical Internet and hyperconnected logistics promise a more efficient and environment-friendly supply chain management, and the Internet of Things as well as blockchain technology appear to be major enablers of such paradigms, especially when it comes to tracking goods throughout the supply chain. On one hand, the IoT is expected to bring a faster goods handling, less prone to processing errors and intentional misbehaviors compared to humans, and a continuous control of the goods condition through various sensors. Blockchain, on the other hand, provides a secure and shared database system for shipments' data provided by IoT devices as well as human operators.

By improving an existing AnyLogic simulation, we showed how blockchain could be concretely applied to hyperconnected logistics, using a private Ethereum network and specific smart contracts. Moreover, we were able to monitor actions stored in blockchain transactions, using Ethereum log events as input trace for our BeepBeep monitor. This, in turn, allowed us to verify basic and more sophisticated parcel lifecycle properties. Such verification is essential. Indeed, although a blockchain backend admittedly enforces an authenticated and immutable database, it does not completely eradicate threats of thefts or failures (e.g., a malicious actor or compromised sensor could still send falsified data, or send no data at all). In this work, the blockchain rather serves as a collaborative record which can be used to monitor security and lifecycle properties in real-time, so that anomalies may be discovered as quickly as possible.

Implementation results showed that performance of such property monitoring reasonably matches actual logistics' scenarios, and that blockchain platforms, with their relatively low transaction rate, are most likely to be the bottleneck in this context. However, recent developments in this area—especially based on new consensus mechanisms used for block generation—hold the promise that blockchain platforms will soon be able to scale up to logistics and supply chain requirements, while being significantly less energy-consuming and just as secure.

References

1. Montreuil, B.: Toward a physical internet: meeting the global logistics sustainability grand challenge. Logist. Res. **3**(2–3), 71–87 (2011). https://doi.org/10.1007/s12159-011-0045-x
2. Crainic, T.G., Montreuil, B.: Physical internet enabled hyperconnected city logistics. Transp. Res. Procedia **12**, 383–398 (2016). https://doi.org/10.1016/j.trpro.2016.02.074
3. Tian, F.: A supply chain traceability system for food safety based on HACCP, blockchain & Internet of things. In: 2017 International Conference on Service Systems and Service Management, pp. 1–6. IEEE, Dalian, China (2017). https://doi.org/10.1109/ICSSSM.2017.7996119
4. Hallé, S., Khoury, R., Betti, Q., El-Hokayem, A., Falcone, Y.: Decentralized enforcement of document lifecycle constraints. Inf. Syst. **74**(Part), 117–135 (2018). https://doi.org/10.1016/j.is.2017.08.002
5. Korpela, K., Hallikas, J., Dahlberg, T.: Digital supply chain transformation toward blockchain integration. In: Bui, T. (ed.) 50th Hawaii International Conference on System Sciences, HICSS 2017, Hilton Waikoloa Village, Hawaii, USA, January 4-7, 2017, pp. 1–10. ScholarSpace / AIS Electronic Library (AISeL) (2017). https://doi.org/10.24251/HICSS.2017.506
6. Lu, Q., Xu, X.: Adaptable blockchain-based systems: a case study for product traceability. IEEE Softw. **34**(6), 21–27 (2017). https://doi.org/10.1109/MS.2017.4121227

7. Madhwal, Y., Panfilov, P.: Industrial Case: Blockchain on Aircraft's Parts Supply Chain Management. In: American Conference on Information Systems (AMCIS), vol. 6. AIS Electronic Library (AISeL) (2017). https://aisel.aisnet.org/sigbd2017/6
8. Mo, L.: A study on modern agricultural products logistics supply chain management mode based on IOT. In: Second International Conference on Digital Manufacturing and Automation, ICDMA 2011, Zhangjiajie, Hunan, China, August 5–7, 2011, pp. 117–120. IEEE Computer Society (2011). https://doi.org/10.1109/ICDMA.2011.36
9. Toyoda, K., Mathiopoulos, P.T., Sasase, I., Ohtsuki, T.: A novel blockchain-based product ownership management system (POMS) for anti-counterfeits in the post supply chain. IEEE Access **5**, 17465–17477 (2017). https://doi.org/10.1109/ACCESS.2017.2720760
10. Wang, X., Liu, N.: The application of internet of things in agricultural means of production supply chain management. J. Chem. Pharm. Res. **6**(7), 2304–2310 (2014). http://www.jocpr.com/articles/the-application-of-internet-of-things-in-agricultural-means-of-production-supply-chain-management.pdf
11. Betti, Q., Khoury, R., Hallé, S., Montreuil, B.: Improving hyperconnected logistics with blockchains and smart contracts. IT Prof. **21**(4), 25–32 (2019). https://doi.org/10.1109/MITP.2019.2912135
12. Kaboudvand, S., Montreuil, B., Buckley, S., Faugere, L.: Hyperconnected Megacity Logistics Service Network Assessment: A Simulation Sandbox Approach (2018). 2018 IISE Annual Conference
13. Company, T.A.: AnyLogic: Simulation Modeling Software Tools & Solutions for Business. https://www.anylogic.com/. Accessed 20 July 2019
14. Hallé, S.: Event Stream Processing with BeepBeep 3: Log Crunching and Analysis Made Easy. Presses de l'Université du Québec (2018)
15. Montreuil, B., Buckley, S., Faugere, L., Khir, R., Derhami, S.: Urban Parcel Logistics Hub and Network Design: The Impact of Modularity and Hyperconnectivity. In: 15th IMHRC Proceedings, vol. 19. Savannah, Georgia. USA (2018). https://digitalcommons.georgiasouthern.edu/pmhr_2018/19
16. Montreuil, B.: The Physical Internet: A Conceptual Journey (2015). Keynote speech. 2nd International Physical Internet Conference. Paris, France, July 6–8, 2015
17. Montreuil, B.: Radical Changes in Freight Transport Systems (2018). Keynote speech. 5th International Workshop on Sustainable Road Freight Transportation. Cambridge, United Kingdom, November 29–30, 2018
18. Rifkin, J.: The Zero Marginal Cost Society: The Internet of Things, the Collaborative Commons, and the Eclipse of Capitalism. St. Martin's Press, New York, NY (2014)
19. Nakamoto, S.: Bitcoin: A Peer-to-Peer Electronic Cash System. Tech. rep. (2008). https://bitcoin.org/bitcoin.pdf
20. Munoz, D., Constantinescu, D., Asenjo, R., Fuentes, L.: Clinicappchain: A low-cost blockchain hyperledger solution for healthcare. In: Prieto et al. [60], pp. 36–44. https://doi.org/10.1007/978-3-030-23813-1_5
21. Odelu, V.: IMBUA: identity management on blockchain for biometrics-based user authentication. In: Prieto et al. [60], pp. 1–10. https://doi.org/10.1007/978-3-030-23813-1_1
22. Mendling, J., Weber, I., van der Aalst, W.M.P., vom Brocke, J., Cabanillas, C., Daniel, F., Debois, S., Ciccio, C.D., Dumas, M., Dustdar, S., Gal, A., García-Bañuelos, L., Governatori, G., Hull, R., Rosa, M.L., Leopold, H., Leymann, F., Recker, J., Reichert, M., Reijers, H.A., Rinderle-Ma, S., Solti, A., Rosemann, M., Schulte, S., Singh, M.P., Slaats, T., Staples, M., Weber, B., Weidlich, M., Weske, M., Xu, X., Zhu, L.: Blockchains for business process management - challenges and opportunities. ACM Trans. Management Inf. Syst. **9**(1), 4:1–4:16 (2018). https://doi.org/10.1145/3183367
23. Kshetri, N.: Can blockchain strengthen the internet of things? IT Prof. **19**(4), 68–72 (2017). https://doi.org/10.1109/MITP.2017.3051335
24. Szabo, N.: Smart Contracts: Building Blocks for Digital Markets. Extropy: J. Transhumanist Thought **16** (1996). http://www.alamut.com/subj/economics/nick_szabo/smartContracts.html

25. Szabo, N.: Formalizing and Securing Relationships on Public Networks. First Monday **2**(9) (1997). https://doi.org/10.5210/fm.v2i9.548
26. Foundation, E.: Solidity 0.5.10 documentation. https://solidity.readthedocs.io/en/v0.5.10/. Accessed 20 July 2019
27. Yardley, J., Barboza, D.: Despite Warnings, China's Regulators Failed to Stop Tainted Milk. The New York Times (2008). URL https://www.nytimes.com/2008/09/27/world/asia/27milk.html
28. Abeyratne, S.A., Monfared, R.P.: Blockchain ready manufacturing supply chain using distributed ledger. Int. J. Res. Eng. Technol. **5**(9), 1–10 (2016). https://doi.org/10.15623/ijret.2016.0509001
29. Kshetri, N.: 1 blockchain's roles in meeting key supply chain management objectives. Int J. Inf. Manag. **39**, 80–89 (2018). https://doi.org/10.1016/j.ijinfomgt.2017.12.005
30. Wang, Y., Han, J.H., Beynon-Davies, P.: Understanding blockchain technology for future supply chains: a systematic literature review and research agenda. Supply Chain. Manag.: Int. J. **24**(1), 62–84 (2019). https://doi.org/10.1108/SCM-03-2018-0148
31. BigchainDB GmbH: BigchainDB 2.0 The Blockchain Database. Tech. rep., Berlin, Germany (2018). https://www.bigchaindb.com/whitepaper/bigchaindb-whitepaper.pdf
32. Shanahan, C., Kernan, B., Ayalew, G., McDonnell, K., Butler, F., Ward, S.: A framework for beef traceability from farm to slaughter using global standards: an Irish perspective. Comput. Electron. Agric. **66**(1), 62–69 (2009). https://doi.org/10.1016/j.compag.2008.12.002
33. Web3 Labs: web3j - Lightweight Ethereum Java and Android integration library. https://web3j.io/. Accessed 20 July 2019
34. Web3 Labs: Solidity smart contract wrappers – web3j. https://docs.web3j.io/smart_contracts/#solidity-smart-contract-wrappers. Accessed 20 July 2019
35. de Morais, R.M., Kazan, S., de Pádua, S.I.D., Costa, A.L.: An analysis of BPM lifecycles: from a literature review to a framework proposal. Bus. Proc. Manag. J. **20**(3), 412–432 (2014). https://doi.org/10.1108/BPMJ-03-2013-0035
36. Falcone, Y., Nazarpour, H., Jaber, M., Bozga, M., Bensalem, S.: Tracing distributed component-based systems, a brief overview. In: Colombo and Leucker [15], pp. 417–425. https://doi.org/10.1007/978-3-030-03769-724
37. Bauer, A., Falcone, Y.: Decentralised LTL monitoring. Form. Methods Syst. Des. **48**(1–2), 46–93 (2016). https://doi.org/10.1007/s10703-016-0253-8
38. Bonakdarpour, B., Fraigniaud, P., Rajsbaum, S., Travers, C.: Challenges in fault-tolerant distributed runtime verification. In: Margaria and Steffen [48], pp. 363–370. https://doi.org/10.1007/978-3-319-47169-3_27
39. El-Hokayem, A., Falcone, Y.: Monitoring decentralized specifications. In: Bultan, T., Sen K. (eds.) Proceedings of the 26th ACM SIGSOFT International Symposium on Software Testing and Analysis, Santa Barbara, CA, USA, July 10–14, 2017, pp. 125–135. ACM (2017). https://doi.org/10.1145/3092703.3092723
40. Falcone, Y., Cornebize, T., Fernandez, J.: Efficient and generalized decentralized monitoring of regular languages. In: Ábrahám, E., Palamidessi, C. (eds.) Formal Techniques for Distributed Objects, Components, and Systems—34th IFIP WG 6.1 International Conference, FORTE 2014, Held as Part of the 9th International Federated Conference on Distributed Computing Techniques, DisCoTec 2014, Berlin, Germany, June 3–5, 2014. Proceedings, *Lecture Notes in Computer Science*, vol. 8461, pp. 66–83. Springer (2014). https://doi.org/10.1007/978-3-662-43613-4_5
41. Sen, K., Vardhan, A., Agha, G., Rosu, G.: Efficient decentralized monitoring of safety in distributed systems. In: Finkelstein, A., Estublier, J., Rosenblum D.S. (eds.) 26th International Conference on Software Engineering (ICSE 2004), 23–28 May 2004, Edinburgh, United Kingdom, pp. 418–427. IEEE Computer Society (2004). https://doi.org/10.1109/ICSE.2004.1317464
42. Ellul, J., Pace, G.J.: Runtime verification of ethereum smart contracts. In: 14th European Dependable Computing Conference, EDCC 2018, Iaşi, Romania, September 10–14, 2018, pp. 158–163. IEEE Computer Society (2018). https://doi.org/10.1109/EDCC.2018.00036

43. Hallé, S., Gaboury, S., Bouchard, B.: Activity recognition through complex event processing: First findings. In: Bouchard, B., Giroux, S., Bouzouane, A., Gaboury, S. (eds.) Artificial Intelligence Applied to Assistive Technologies and Smart Environments, Papers from the 2016 AAAI Workshop, Phoenix, Arizona, USA, February 12, 2016, *AAAI Workshops*, vol. WS-16-01. AAAI Press (2016)
44. Jia, X., Wenming, Y., Dong, W.: Complex event processing model for distributed RFID network. In: Sohn, S., Chen, L., Hwang, S., Cho, K., Kawata, S., Um, K., Ko, F.I.S., Kwack, K., Lee, J.H., Kou, G., Nakamura, K., Fong, A.C.M., Ma, P.C.M. (eds.) Proceedings of the 2nd International Conference on Interaction Sciences: Information Technology, Culture and Human 2009, Seoul, Korea, 24–26 November 2009, *ACM International Conference Proceeding Series*, vol. 403, pp. 1219–1222. ACM (2009). https://doi.org/10.1145/1655925.1656147
45. Wang, F., Zhou, C., Nie, Y.: Event processing in sensor streams. In: C.C. Aggarwal (ed.) Managing and Mining Sensor Data, pp. 77–102. Springer (2013). https://doi.org/10.1007/978-1-4614-6309-2_4
46. van der Aalst, W.M.P.: Process Mining - Data Science in Action. 2nd edn. Springer (2016). https://doi.org/10.1007/978-3-662-49851-4
47. Calvar, J., Tremblay-Lessard, R., Hallé, S.: A runtime monitoring framework for event streams with non-primitive arguments. In: Antoniol, G., Bertolino, A., Labiche Y. (eds.) Fifth IEEE International Conference on Software Testing, Verification and Validation, ICST 2012, Montreal, QC, Canada, April 17–21, 2012, pp. 499–508. IEEE Computer Society (2012). https://doi.org/10.1109/ICST.2012.135
48. Jin, D., Meredith, P.O., Lee, C., Rosu, G.: Javamop: Efficient parametric runtime monitoring framework. In: Glinz, M., Murphy, G.C. Pezzè, M. (eds.) 34th International Conference on Software Engineering, ICSE 2012, June 2–9, 2012, Zurich, Switzerland, pp. 1427–1430. IEEE Computer Society (2012). https://doi.org/10.1109/ICSE.2012.6227231
49. Leucker, M., Schallhart, C.: A brief account of runtime verification. J. Log. Algebr. Program. **78**(5), 293–303 (2009). https://doi.org/10.1016/j.jlap.2008.08.004
50. Adi, A., Botzer, D., Nechushtai, G., Sharon, G.: Complex event processing for financial services. In: Proceedings of the 2006 IEEE Services Computing Workshops (SCW 2006), 18–22 September 2006, Chicago, Illinois, USA, pp. 7–12. IEEE Computer Society (2006). https://doi.org/10.1109/SCW.2006.7
51. Berry, A., Milosevic, Z.: Real-time analytics for legacy data streams in health: Monitoring health data quality. In: Gasevic, D., Hatala, M., Nezhad, H.R.M., Reichert M. (eds.) 17th IEEE International Enterprise Distributed Object Computing Conference, EDOC 2013, Vancouver, BC, Canada, September 9–13, 2013, pp. 91–100. IEEE Computer Society (2013). https://doi.org/10.1109/EDOC.2013.19
52. La, V.H., Fuentes-Samaniego, R.A., Cavalli, A.R.: Network monitoring using MMT: an application based on the user-agent field in HTTP headers. In: Barolli, L., Takizawa, M., Enokido, T., Jara, A.J., Bocchi, Y. (eds.) 30th IEEE International Conference on Advanced Information Networking and Applications, AINA 2016, Crans-Montana, Switzerland, 23–25 March, 2016, pp. 147–154. IEEE Computer Society (2016). https://doi.org/10.1109/AINA.2016.41
53. Luckham, D.C.: The power of events—an introduction to complex event processing in distributed enterprise systems. ACM (2005)
54. Boussaha, M.R., Khoury, R., Hallé, S.: Monitoring of security properties using BeepBeep. In: Imine, A., Fernandez, J.M., Marion, J., Logrippo, L., García-Alfaro J. (eds.) Foundations and Practice of Security - 10th International Symposium, FPS 2017, Nancy, France, October 23–25, 2017, Revised Selected Papers, *Lecture Notes in Computer Science*, vol. 10723, pp. 160–169. Springer (2017). https://doi.org/10.1007/978-3-319-75650-9_11
55. Hallé, S., Gaboury, S., Bouchard, B.: Towards user activity recognition through energy usage analysis and complex event processing. In: Proceedings of the 9th ACM International Conference on PErvasive Technologies Related to Assistive Environments, PETRA 2016, Corfu Island, Greece, June 29–July 1, 2016, p. 3. ACM (2016). https://doi.org/10.1145/2910674

56. Hallé, S., Gaboury, S., Khoury, R.: A glue language for event stream processing. In: Joshi, J., Karypis, G., Liu, L., Hu, X., Ak, R., Xia, Y., Xu, W., Sato, A., Rachuri, S., Ungar, L.H., Yu, P.S., Govindaraju, R., Suzumura, T. (eds.) 2016 IEEE International Conference on Big Data, BigData 2016, Washington DC, USA, December 5–8, 2016, pp. 2384–2391. IEEE (2016). https://doi.org/10.1109/BigData.2016.7840873. http://ieeexplore.ieee.org/xpl/mostRecentIssue.jsp?punumber=7818133
57. Khoury, R., Hallé, S., Waldmann, O.: Execution trace analysis using LTL-FO+. In: Margaria and Steffen [48], pp. 356–362. https://doi.org/10.1007/978-3-319-47169-3_26
58. Varvaressos, S., Lavoie, K., Gaboury, S., Hallé, S.: Automated bug finding in video games: A case study for runtime monitoring. Comput. Entertain. **15**(1), 1:1–1:28 (2017). https://doi.org/10.1145/2700529
59. Hallé, S., Khoury, R.: Writing domain-specific languages for beepbeep. In: Colombo and Leucker [15], pp. 447–457. https://doi.org/10.1007/978-3-030-03769-7_27
60. Hallé, S.: When RV meets CEP. In: Falcone, Y., Sánchez, C. (eds.) Runtime Verification - 16th International Conference, RV 2016, Madrid, Spain, September 23–30, 2016, Proceedings, *Lecture Notes in Computer Science*, vol. 10012, pp. 68–91. Springer (2016). https://doi.org/10.1007/978-3-319-46982-9_6
61. Moore Edward F.: Gedanken-Experiments on Sequential Machines. Automata Studies (AM-34) **34** (1956). https://doi.org/10.1515/9781400882618-006
62. Roudjane, M., Rebaine, D., Khoury, R., Hallé, S.: Real-time data mining for event streams. In: 22nd IEEE International Enterprise Distributed Object Computing Conference, EDOC 2018, Stockholm, Sweden, October 16–19, 2018, pp. 123–134. IEEE Computer Society (2018). https://doi.org/10.1109/EDOC.2018.00025
63. Coin Dance: Bitcoin Nodes Summary. https://coin.dance/. Accessed: 2019-07-20
64. ethernodes.org: The ethereum node explorer. https://www.ethernodes.org/network/1. Accessed: 20 July 2019
65. Zhang, Y., Kasahara, S., Shen, Y., Jiang, X., Wan, J.: Smart contract-based access control for the internet of things. IEEE Internet Things J. **6**(2), 1594–1605 (2019). https://doi.org/10.1109/JIOT.2018.2847705
66. Visa: Annual Report (2018). https://s1.q4cdn.com/050606653/files/doc_financials/annual/2018/Visa-2018-Annual-Report-FINAL.pdf. Accessed 20 July 2019
67. Block.one: EOS.IO Technical White Paper v2. Tech. rep. (2018). https://github.com/EOSIO/Documentation
68. CryptoLions: EOS Network Monitor. https://eosnetworkmonitor.io/. Accessed: 2019-07-20
69. Futurepia: Futurepia Mainnet, Blockchain for Social Media. https://futurepia.io/. Accessed 20 July 2019
70. Bitshares: Delegated Proof-of-Stake Consensus. https://bitshares.org/. Accessed: 2019-07-20
71. Futurepia: Futurepia: Built and Owned by User, White Paper Ver. 1.3. Tech. rep. (2019). https://futurepia.io/assets/img/FUTUREPIA_WhitePaper_EN.pdf
72. Betti, Q., Hallé, S.: Benchmark for supply chain monitoring properties using BeepBeep (LabPal dataset) (2019). https://doi.org/10.5281/zenodo.3066013
73. Hallé, S., Khoury, R., Awesso, M.: Streamlining the inclusion of computer experiments in a research paper. IEEE Comput. **51**(11), 78–89 (2018). https://doi.org/10.1109/MC.2018.2876075

Validating BGP Update Using Blockchain-Based Infrastructure

Kolade Folayemi Awe, Yasir Malik, Pavol Zavarsky and Fehmi Jaafar

Abstract A number of solutions have been proposed to secure the Border Gateway Routing (BGP) protocol by validating BGP update path and origin information. These solutions make use of centralized database, centralized Public Key Infrastructure (PKI) and some conventional PGP variants as their security mechanism. These solutions are prone to successful attack by state actors and often build database to verifying BGP updates without proper means of validating data stored in this database. Therefore, there is a need for alternative approach to secure the BGP routing protocol. In this chapter, we propose a blockchain based technology used to create a distributed or decentralized immutable database that relies on consensus of participating Autonomous System (AS), to build this blockchain. Every BGP route update received by an AS peer is validated against the content of the blockchain distributed database to detect updates with falsified path and origin information. The limitation of throughput and scalability associated with the blockchain would not affect the proposed blockchain solution once it is fully operational. This is because the data stored in the distributed ledger has a frequency or rate of change that is far lower than that of the blockchain transaction rate. Furthermore, with the blockchain solution, the centralized PKI root of trust is eliminated and AS are now capable of detecting and mitigating IP prefix hijack attack in real time, without outsourcing this service to a third party.

Keywords The border gateway routing · The public key infrastructure · Blockchain based technology

K. F. Awe · Y. Malik · P. Zavarsky · F. Jaafar
Concordia University Edmonton, Alberta, Canada
e-mail: kawe@csa.concordia.ab.ca

Y. Malik
e-mail: yasir.malik@concordia.ab.ca

P. Zavarsky
e-mail: pavol.zavarsky@concordia.ab.ca

F. Jaafar (✉)
The Computer Research Institute, Quebec, Canada
e-mail: fehmi.jaafar@crim.ca

M. A. Khan et al. (eds.), *Decentralised Internet of Things*, Studies in Big Data 71,
https://doi.org/10.1007/978-3-030-38677-1_7

1 Introduction

The Border Gateway Protocol (BGP), described in RFC 1771 is used to exchange IP prefix reachability information among different Autonomous System (AS) ensuring operation of the Internet. In view of this, the BGP is the only routing protocol in use on the Internet today.

Indeed, the BGP routing protocol basically provides reachability information for different Autonomous System (AS) across the Internet. An autonomous system is a group of nodes or infrastructure under the same administrative control. Autonomous systems propagate their IP prefix and AS number to the rest of the Internet. BGP routing protocol takes the responsibility of ensuring all ASs can reach each other. Thus, autonomous systems or sub-networks would be isolated or unreachable without the BGP protocol [1]. BGP provides many attributes to support policy routing based on contracts between independent ASs operators for connectivity to the Internet of Things. There are two major types of BGP routing protocol: the iBGP (interior Border Gateway Protocol) and eBGP (exterior Border Gateway Protocol). The iBGP is used to route traffic within an autonomous system, while the eBGP interconnects Autonomous Systems (AS) across the Internet.

One of the main challenge regarding the use of the BGP protocol in the Internet of Things, is to improve its security posture. In fact, the BGP protocol was developed in the 1980s with little or no security mechanism built in it, to meet modern day threat across the Internet. It is obvious today, that the Internet is an insecure and uncontrolled environment, where malicious users exploit every possible means to execute an attack. The vulnerabilities in the Border Gateway routing protocol have been exploited by attackers to perpetrate their malicious intent. For example, the BGP has been used to manipulate routing traffic direction or announce incorrect update by malicious users.

In this chapter, we propose a distributed, immutable and peer to peer database system that is not controlled by any central body to mitigate BGP IP prefix hijack on the Internet. Leveraging on the technology of existing cryptocurrency, where two willing parties can transact together without the need for a trusted third party. In this system, transactions recorded into the blockchain cannot be reversed. This solution equips each AS with a database that allows external BGP speakers within the AS validate the authenticity of IP prefix origin and AS path information contained in every BGP update received. The blockchain technology allows peers that would otherwise not trust each other to transact together, because they trust the mechanisms that drive the infrastructure. In this chapter, the blockchain distributed ledger is proposed to secure the BGP routing protocol from IP prefix hijack attack. Though the blockchain was designed as the core technology that drives crypto currency transactions, we describe a cryptocurrency like blockchain solution to secure or mitigate BGP IP prefix hijack attack. Similar to how transactions are verified, stored and are immutable in the cryptocurrency blockchain, hence preventing misuse, BGP data stored in the proposed blockchain solution would also be verified and be immutable [2]. The blockchain solution proposed to secure the BGP routing protocol from IP prefix hijack attack (by validating BGP route origin and path information) eliminates the need for central certificate authorities (CA) and web of trust approach used by the other approaches proposed to secure the BGP. Therefore, using a blockchain makes the proposed solution and infrastructure fully decentralized. The blockchain solution proposed would inherit existing security features of the cryptocurrency blockchain such as proof of work, wallet, and mining nodes [3]. The goal of this chapter is not to propose a new routing protocol that mitigates IP prefix hijack attack initiated by manipulating route origin or AS path. Instead, we propose a decentralized, immutable database that validates received BGP route updates to detect and mitigate updates with falsified route origin and AS path information. Once an update is identified to have falsified route origin or AS path information, it is not stored or used by a BGP speaking peer.

The rest of this chapter is organized as follows, Sect. 2 presents review of related work on the mitigation of IP prefix hijack, Sect. 3 discusses in detail the components of the proposed blockchain solution. Section 4 illustrates how the different components work in concert with the BGP protocol to mitigate IP prefix hijack attack, by validating IP prefix origin and AS path information contained in each route update. The results of an experiment are presented in Sect. 5. This chapter concludes in Sect. 6.

2 Related Work

The literature review will be divided into two parts. The first part (A) refers to related works used to validate either BGP AS path or BGP IP prefix origin with solutions that are not related to cryptocurrencies or blockchain. The second part (B) reviews related works used to validate BGP AS path and route origin information using solutions that are related to the blockchain technology.

2.1 Autonomous Systems Path Verification

Wang et al. in [4], proposes a new secure inter-domain routing protocol called identity-based inter-domain routing (id2r). id2r is made up of a BGP origin verification mechanism called AT (Assignment Track) and an AS path verification mechanism known as IBAPV (identity-Based Aggregate Path Verification).

The limitation of the id2r is that the ATA repository relies on trust from connected AS. It believes the information provided by its peer AS without any method of authentication. The ATA repository of different servers also exchanges AS to prefix list mappings that are not verified. This information can be manipulated or forged without the receiving AS detecting it. This solution is not compatible with BGP v4 presently in use today and would require the design of a new protocol. This may result in a massive infrastructure overhaul across AS or ISP on the Internet.

In our proposed blockchain solution, an AS does not trust the information provided by directly connected AS, as it can be falsified.

Furthermore, our proposed blockchain solution is compatible with BGP v4 there is no need to change any protocol field information in the IP header.

Vidya et al. in [5], proposed the use of logarithmic keying technique to secure AS-PATH information by combining networks across the Internet. Symmetric keys are used for cryptographic operations in securing AS-PATH by viewing the Internet as a group of acyclic and star networks comprised of ISPs and end users. The flaw in this paper is that large transit AS, called grand-parent AS are trusted to identify falsified AS-Path. The trusted AS has no restriction or control to executing a prefix hijack or AS Path falsification. Symmetric key algorithm has its security limitation of effective key distribution and use. Our proposed blockchain solution in this chapter creates a distributed database utilizing asymmetric keys, comprising of private/public key pair. This eradicates the limitation of symmetric key distribution and other issues related to symmetric keys. The proposed blockchain solution uses a peer to peer network that provides equal security to each AS.

Butler et al. in [6], proposed BGP path authentication mechanism that reduces cryptographic cost. The trade-off for cryptographic cost reduction is built around the hypothesis that BGP announcements to an AS usually follow a set of stable or fixed path across the Internet. Autonomous Systems (AS) were constructed to form a hash tree for path aggregation with AS along an AS-Path referred to as the leaves. The hash of the AS sibling, parents siblings etc up to the root are required along the path. Therefore, fixed cryptographic hash values are computed to authenticate each AS-Path. The limitation identified in this related work is that Secure BGP type route attestation is used on BGP route updates with new paths. Applying cryptographic operation on BGP route updates that is time critical and rapidly changing everyday results in high cryptographic overhead and is computationally expensive. Furthermore, with the expansion of the Internet, the hash tree becomes very large increasing number of hash operations [5]. This solution is not compatible with the present BGP protocol in use on the Internet today. Therefore a massive overhaul of infrastructure would be required if this solution is to be used. The proposed blockchain solution executes cryptographic operation on data that is not rapidly been exchanged or propagated across the Internet, such as new AS to IP prefix assignment and new BGP neighbor peering. This causes the overall cryptographic operation to be minimal on daily basis and in the long run.

The study by Kent et al. in [7] proposed the use of Secure BGP (sBGP) to solve BGP route origin and path authentication vulnerability issues. It uses the public key infrastructure, route/address attestation and IPSEC as its security mechanism. Address attestation is used for origin authentication, while route attestation is used for path authentication. sBGP applies excessive cryptographic operation on BGP route updates that is exchanged very often and is rapidly changing. This makes the proposed solution cryptographically expensive, therefore reducing its feasibility of being deployed. Implementing sBGP also requires the design of a new routing protocol, which would result in a massive overhaul of the Internet routing infrastructure.

The proposed blockchain solution executes cryptographic operation on data that is not rapidly been exchanged or propagated across the Internet, such as new AS to IP prefix assignment and new BGP neighbor peering. This result is relatively low cryptographic operation when considering the cryptographic operation volume over time. The proposed solution is also compatible with the present BGP routing protocol.

Sermpezis et al. in [8], proposes the automatic and Real-Time dEtection and Mitigation System (ARTEMIS). It is capable of detecting an IP prefix hijack incident in real-time and mitigating IP prefix hijack automatically for prefixes that are within their administrative control within two to five minutes. ARTEMIS relies on a database that is constantly updated and the ARTEMIS solution guarantees 0% false positive and false negative information. Therefore, ARTEMIS solution allows a shift from a third party perspective to self-managed solution. ARTEMIS relies on common BGP feeds such as BGPmon RouteView, and RIPE RIS to detect IP prefix hijack. All three monitors use updates on BGP routers all around the world, otherwise called route collectors. However, not all data received from these route collectors are collected in real time. ARTEMIS compares AS-PATH and prefix fields in BGP announcements with the data in the local database (obtained from RIPE RIS, RouteView and BGPmon) to detect a prefix hijack and generate alerts. ARTEMIS limitation is that it relies on a database from route collectors that is not updated in real time. The blockchain solution proposed in this chapter, validates BGP route update, AS path and IP prefix origin information, using a distributed database that is real time and up-to-date.

2.2 *Resolving the BGP Vulnerabilities Using Blockchain*

Adiseshu et al. in [2] proposed the use of the blockchain to verify BGP update origin and detect manipulated AS path information. The proposed solution uses a blockchain converter module to convert outgoing BGP updates into advertisement transactions. These advertisement transactions are then broadcasted on the blockchain peer-to-peer network and is received by Autonomous Systems. The blockchain converter module converts them back from advertisement transactions in the blockchain to BGP updates.

The cryptocurrency scalability and throughput issue allow a block to be added to the blockchain once every 10 minutes. Therefore, limiting transactions rate to about 7 transactions per second. BGP updates received by a peer peaks to about 9000 prefix changes per second [2]. Collecting BGP update information into the blockchain database would result in significant delay that would result in frequent to permanent Internet outage. The proposed blockchain solution in this chapter does not store BGP updates into the blockchain distributed database. Instead, information that is not rapidly changing is stored in the blockchain distributed database such as AS number to IP prefix assignment and new neighbor connection information. This information can be accommodated into the blockchain with the present transaction rate of 7 transaction per second.

de la Rocha et al. in [3] proposes the use of Secure Blockchain Trust Management (SBTM) to complement the BGP routing protocol. The proposed solution uses the blockchain technology to store Autonomous Systems (AS) certificate, which are used for encryption and decryption of BGP secure route update information, when SBGP or SoBGP is used. This solution complements an existing secure routing protocol. Therefore, it is a robust and scalable solution. The SBTM complements a secure inter domain routing protocol; this means that it can operate only when Secure BGP (SBGP) and Secure Origin BGP (SoBGP) are fully deployed on the Internet. SBGP and SoBGP are considered too expensive cryptographically for use on the Internet today. Therefore, it is not in use on the Internet yet. This makes the chances of deploying the SBTM on the Internet low. The combination of SBGP and SoBGPs cryptographic operation with that of the blockchain would be way too expensive cryptographically and computationally. The proposed blockchain solution in this chapter can work with and secure the BGP protocol presently in use on the Internet. Furthermore, the proposed blockchain solution applies cryptographic operations on the blockchain and not BGP routing updates.

The application of cryptographic operation on BGP route updates that is time critical and rapidly changing every day results in high cryptographic overhead and is computationally expensive. Furthermore, related work that applies cryptographic operations to BGP route updates requires massive change in the BGP protocol header information, resulting in the development of a new routing protocol. The blockchain solution proposed in this chapter, does not propose the development of a new BGP routing protocol. It works by complementing the BGP routing protocol deployed on the Internet today, by validating BGP AS path and route origin information received by each AS.

Cryptocurrency scalability refers to throughput limitation related to transaction rate. At present the bitcoin has a transaction rate of 7tps (transactions per second), while transaction platforms such as VISA has 2000 tps and Twitter has 5000 tps. Improving the frequency of bitcoin transactions is an area where intense research is ongoing at the moment [9]. In view of this statistic, an attempt to store BGP update messages having a peak rate of 9000 prefix changes/s in the blockchain is not feasible for now [2]. The solution proposed in this chapter, stores information such as new AS number to IP prefix allocation and new AS neighbor connection in the blockchain. These information have a change rate that can be well accommodated within 7tps. A number of related works that secure the BGP routing protocol by authentication or validating AS path or routing origin information, assign trust status to specific number of AS. These trusted AS are capable of manipulating route updates without any checks or validation. AS do not trust each other in the proposed blockchain solution. They however trust the blockchain security mechanism used to drive the communication. The proposed blockchain solution eliminates the need for a trusted Certificate Authority (CA). The next section presents the components and processes required by each AS to participate in the proposed blockchain solution.

3 Blockchain Components

In the cryptocurrency world, the blockchain is a distributed database that keeps a list of records that is growing persistently, these list of records are called blocks. The blockchain is used by a peer to peer network that act in accordance with the mechanism or protocol that validates new blocks. The blockchain is designed such that data stored into it cannot be modified. Modification of the data already stored in a single block would require the alteration of subsequent blocks and collusion with several nodes on network [10]. Our proposed blockchain solution would adopt the same components, and processes as the cryptocurrency blockchain. These components and processes would be required for an AS to participate in the blockchain and initiate a transaction. The next subsection discusses these components and processes and their relationship with the proposed blockchain solution.

3.1 Wallet, Keys and Multi-signature Keys

In cryptocurrency, the wallet stores identities and cryptographic keys required by users to participate in the blockchain protocol. It also gives each user the right to inspect and participate in the blockchain. A single signature key pair comprising of a private and public key is used for standard transactions in the bitcoin network. The multi-signature key is a more secure option and can also be used [11]. Each AS participating in the proposed blockchain solution adopts and uses the more secure multisig wallet that stores the multi-signature keys. The multisig wallet and multi-signature keys are required to participate in the blockchain and initiate transactions.

3.2 Transactions

A sequence of information or messages represents what is called a transaction in the cryptocurrency world. A bitcoin transaction is made up of two parts: an input and output [12]. The proposed blockchain solution uses the same concept for transactions. Each transaction initiated by an AS or the RIR consist of the input part and the output part. The output part of the transaction can be redeemed by the AS with the appropriate private key. More details of transactions would be given in Sect. 4 of this chapter.

3.3 Mining

In the cryptocurrency world, the mining operation is done by computers and their unit for measurement is hashes per second. The miners also gets as a reward a certain number of bitcoins per block [11]. The Proof of Work can be described as the process of searching, scanning or guessing for a value such that when this value is hashed, the result of the hash operation begins with a number of leading zero bits. The work that needs to be done is usually an exponential number of zero bits and the result can be verified by calculating a single hash [13]. The proposed blockchain solution would leverage on the same mining mechanism and concept used in the cryptocurrency world.

3.4 *The Content of the Blockchain*

In the cryptocurrency world, the first block of a blockchain is usually called the genesis block. Each block in the cryptocurrency world contains a value for a timestamp, index, previous hash, a new hash and data. The previous hash value in each block refers to the hash of the prior block in the blockchain, therefore linking both blocks together. A new hash value is generated for each block that links and serves as input to the preceding block in the blockchain. As a proof of concept, we simulate a cryptocurrency blockchain comprising of three blocks linked together. A block function was created with index, timestamp, data, hash and previous hash values. Hash values were calculated and a blockchain function was created with blocks imported into it. Hash values for previous blocks were linked to new blocks. A validation algorithm was embedded into the simulation environment to validate newly created blocks. The data stored in the blockchain distributed ledger is an AS number with its respective IP prefix assigned by the Regional Internet Registry (RIR). The output of this experiment is shown in Fig. 1.

The next section presents details on how the proposed blockchain solution works. How the validation process of BGP updates occur and detailed illustration of the proposed blockchain solution are also discussed.

```
[ Block {
    index: 0,
    previousHash: '0',
    timestamp: 'Tue, 05 Jun 2018 23:08:50 GMT',
    data: 'I am the genosis block',
    hash: 'c94e6f34fadf9e1ba5f43f25796a5fda308b1853e60cc73fc51a27f0bcaba2c7' },
  Block {
    index: 1,
    previousHash: 'c94e6f34fadf9e1ba5f43f25796a5fda308b1853e60cc73fc51a27f0bcaba2c7',
    timestamp: 'Tue, 05 Jun 2018 23:08:50 GMT',
    data: 'AS NO:65101 IP PREFIX 192.168.20.5',
    hash: 'dc5b5dff8e8c7e53b7f96d87633bef2baeb83c55bbe15b1c0d2611d0fab97a4a' },
  Block {
    index: 2,
    previousHash: 'dc5b5dff8e8c7e53b7f96d87633bef2baeb83c55bbe15b1c0d2611d0fab97a4a',
    timestamp: 'Tue, 05 Jun 2018 23:08:50 GMT',
    data: 'AS NO:65102 IP PREFIX 192.168.33.45',
    hash: '49fde3a38d4959a6e98baeea61ffab0f9f96c3d14e742bd54849ae876286bb47' } ]
```

Fig. 1 Contents of each block in a cryptocurrency blockchain

4 BGP AS Path and IP Prefix Validation Using the Blockchain

The proposed blockchain solution is a distributed database that contains information that is logically divided into two parts. The first part validates the authenticity of each BGP announcements route origin information. The BGP route announcement is then passed to the second part of the distributed database for further verification. The second part of the blockchain validates the authenticity of the BGP announcements AS path information. If the result of either or both verification process is false, the announcement is considered an attempted IP prefix hijack attack. It is logged and the BGP route announcement is neither stored into the peers routing table nor used. It is dropped and logged. This gives each participating AS the capability to detect and mitigate an IP prefix hijack attack both independently and in real time. The requirements an autonomous system needs to participate in the proposed blockchain are highlighted below.

4.1 *Autonomous System (AS) Requirements*

- Every AS should be capable of generating and storing its private-public key pair. This information is hashed to form the AS address and is used to grant each AS access to the blockchain.
- Every AS border router that receives an announcement must be capable of validating this announcement against the content of the blockchain. This would decide if the announcement has been manipulated and can result in a prefix hijack or not.
- A new AS that wants to join the BGP blockchain system is required to carry out a blockchain transaction. The generation of its cryptographic keys must precede the transaction.
- The BGP blockchain must provide a security solution that makes an AS capable of detecting and mitigating route announcement with hijacked IP prefix.
- An AS must be recognized by its regional Internet registry and must have been assigned an IP address by it.

4.2 *The Distributed Ledger Database*

The proposed blockchain solution is a distributed ledger that contains information that is divided into two parts. The first part contains AS numbers to IP prefix mappings for all AS number assigned by the RIR. This database authenticates every BGP route update received by an AS to detect and mitigate IP Prefix hijack attack containing manipulated AS origin information. Table 1 shows sample of the first part of the blockchain solution.

Table 1 The first part of the blockchain

AS number	IP prefix
65101	172.20.10.0/24
65102	172.20.20.0/24
65103	172.20.30.0/24
65104	172.20.40.0/24
65105	172.20.50.0/24
65106	172.20.60.0/24
65107	172.20.70.0/24

Table 2 The second part of the blockchain BGP

AS number	Directly connected neighbors
65101	65104
65102	65104, 65105
65103	65104, 65106
65104	65101, 65102, 65103
65105	65102, 65107
65106	65103
65107	65105

The second part of the blockchain shown in Table 2 contains the list of all autonomous system (AS) numbers on the Internet and their directly connected neighbor AS. This part is used to detect IP prefix hijack attack that is executed by manipulating the AS-PATH attribute.

4.3 Mining Operation

Transactions are verified and recorded in a block to create the database shown in Tables 1 and 2, by a process called mining. A transaction initiated by an AS or the Regional Internet Registry (RIR), is eventually stored into a block and broadcasted to the peer to peer AS network for mining. An AS attaches this block in a linear chain called blockchain. Every AS participating in the BGP blockchain is involved in the mining process. The blockchain transaction verification process is achieved by ensuring that each blocks input points to the previous block output. Mining is the responsibility of either the border router or a server in the AS. The Regional Internet Registries (RIR) and ICANN are also involved in the mining process. The mining process validates new transaction that eventually find their way into the blockchain. The mining process tries to solve a complex computational problem that is very processor intensive. The mining process for part 1 and 2 of the blockchain is explained next.

Mining data into the first part of the blockchain is done by the RIR, ICANN and all AS. We briefly outline the transaction process by the RIR and an AS called ASx. The RIR initiates a transaction like that of the cryptocurrency. ASxs AS number and its assigned IP prefix is encrypted using the public key of ASx; this serves as the transaction output. The input to this transaction contains the address of the RIR. This transaction is then encrypted with the private key of the RIR. The transaction is broadcasted out to the entire AS on the Internet or peer to peer network. All AS that receives this transaction stores the AS number of ASx to its IP prefix mapping temporarily in a local buffer. Only ASx can redeem the transaction because, the transaction output is encrypted with its public key and only its private key can decrypt it. ASx redeems the transaction, stores it in a block and broadcast the block into the peer to peer network. All AS receives this block, they then verify the transaction, by (1) validating that the entry corresponds to the AS to IP prefix mapping stored in their buffer and (2) that the IP prefix to AS mapping does not exist in the blockchain. Both conditions 1 and 2 must be valid for each AS to send out a confirmation out to the peer to peer network that the entry in the block is valid. If 60% of the AS validates the entry, then the mining operation begins, the AS that solves the puzzle adds the block to the blockchain. If the consensus received for this transaction is not up to 60%, the block is not added to the blockchain.

Mining data into the second part of the blockchain is done differently from the first part. The second part of the blockchain contains entries of AS numbers and their directly connected neighbors. Each AS sends information of its directly connected neighbor, encrypted with both the private key of the originating AS and the public key of the neighbor. An AS with multiple neighbor connection would initiate separate transactions for each neighbor. The transaction output contains the AS number of the directly connected neighbor and it is encrypted with the public key of the neighbor AS. The AS initiating the transaction encrypts its AS number with its private key as the transaction input. This transaction is broadcasted out the peer to peer network and is received by all AS. Each AS stores the neighbor relationship information temporarily in their local buffer. The neighbor AS uses its private key to redeem the transaction. The transaction is written into a block by the neighbor AS and the block is broadcasted to the peer to peer network. Each AS in the peer to peer network verifies that (1) the transaction in the block corresponds to the AS to AS neighbor connection stored in their buffer earlier and (2) the entry does not exist in the blockchain. Both conditions 1 and 2 must be valid for every AS to send out a confirmation out to the peer to peer network that the entry in the block is valid. If 60% of the AS validates the entry, then the mining operation begins, the AS that solves the puzzle adds the block to the blockchain. If the consensus received for this transaction is not up to 60%, the block is not added to the blockchain.

4.4 The Smart Contract

The Smart Contract are codes or programs that are stored in the blockchain. They are contractual agreement that automate different forms of payments in the cryptocurrency blockchain.

The smart contract codes are executed based on the terms of events that are predetermined. These codes are entered in the blockchain and they operate by executing instructions to carry out specific tasks [14]. When a route update is received by an AS peer, this update is sent to and received by the blockchain module for verification. The codes of the smart contract compare this route update with the content of the distributed ledger or blockchain. Therefore, the smart contract plays an important role in the route origin and AS Path verification using the data stored in the blockchain.

Table 1 is used to validate the route origin for route announcement “a” and “b” received by an AS as shown below:

(a) 65104 → 65103 → 65107–172.20.10.0/24
(b) 65104 → 65103 → 65107–172.20.70.0/24

The IP prefix 172.20.10.0/24 in route announcement “a” originates from AS65107, when validated against the entry in part 1 of the blockchain database; it is clear that it has a manipulated route origin. The original owner of the IP prefix 172.20.10.0/24 in Table is AS65101. The result is logged, and the entry is not added into the BGP peers routing table for use. It is dropped and logged. The IP prefix in the route announcement “b” 172.20.70.0/24 is originated by AS65107. When validated against part 1 of the blockchain entry in Table 1, the route announcement is valid because AS65107 was originally assigned the IP prefix 172.20.70.0/24. The route announcement is passed on for further verification in the second part of the blockchain. The smart contract runs on the distributed database on each AS and verifies the AS-PATH information of every route announcement received. An IP prefix hijack attack with manipulated path information is therefore easily detected and mitigated. When AS65101 receives a route update with manipulated AS Path information, as shown in “c” below:

(c) 65104 → 65103 → 65107–172.20.70.0/24
(d) 65104 → 65102 → 65105 → 65107–172.20.70.0/24

Smart contract performs computation on part two of the blockchain to validate that each hop of received route update “c” and “d” has a direct neighbor relationship with each other. The route announcement “c” above is validated using part two of the blockchain database. AS65104 and AS65103 have direct neighbor connection to each other, hence it is a valid path. AS65103 and AS65107 does not have direct neighbor connection to each other as claimed by the route announcement. Therefore, it has a manipulated AS Path information. This route announcement would not be added into the routing table of the BGP peer for use. Instead it is logged and dropped. Verification of route announcement “d” with part 2 of the blockchain shows that AS65104 has neighbor relationship with AS65102 and AS65102 also has neighbor relationship with AS65105, and AS65105 has neighbor relationship with AS65107. Therefore, route announcement "d" has a valid AS path information. It is added to the BGP peers routing table for use.

The proposed blockchain solution would leverage on existing security functions of the cryptocurrency blockchain such as proof of work, which makes the information stored in the blockchain is almost impossible to change. To Change or manipulate information registered in a blockchain, an attacker will need to modify the content of all prior blockchain. The attacker would require so much computational power and would also have to take control of over 60% of the peer to peer AS network to get their approval. In the next section of this chapter, we conduct an experiment to determine the maximum acceptable blocksize of each block in the blockchain.

5 Experiment

In the blockchain solution, the maximum block size is directly proportional to the volume or number of transactions contained in a block. Stale block also known as orphan block materializes when two miners generate identical blocks at the same time. The inclusion of a mined block into a blockchain by AS does not occur immediately, therefore other miners could solve exactly the same block. AS or nodes would now have to decide which block to build on. The block with the higher proof-of-work wins the tie. The block with smaller proof-of-word is called the stale or orphan block. The frequency of generation of stale blocks is called the stale block rate. High stale block rate weakens the security of the blockchain. The block propagation time grows linearly as the block size increases. Therefore, large block sizes result in slower propagation time. This in turn increases the stale block rate and weakens the overall security of the blockchain [15]. An experiment is conducted showing the block propagation time with an increase in the block size. The block size is measured in megabytes (MB) and the propagation time is measured in seconds (s). The stale block rate in percentile values is shown in Table 3. We used a NS3 based bitcoin simulator to generate a peer to peer network comprising of 40 blocks and 64 nodes. This represents a blockchain with 40 blocks and 64 AS. Rapidjson is used for communication between the nodes. Real network statistics were collected from blockchain.info and integrated into this simulator [16]. Figure 2 shows a graph with average propagation time in seconds on the y-axis and the block size in megabytes (MB) on the x-axis. The average block propagation time increases as the size of the block increases. In the proposed blockchain solution, the maximum acceptable block size of 2 MB is chosen for each block. The average propagation time and the stale rate is at the best value at 2 MB. Furthermore, the maximum acceptable blocksize of 2 MB allows the proposed blockchain solution to scale and meet future data storage requirements. The stale block rate of 8.33% is still within acceptable limits. The average propagation time is shown to be 0.183 s with the maximum block size of 2 MB. Figure 2 also shows that the stale block rate is the lowest at a block size of 2 MB.

Table 3 Experiment displays stale block rate with increase in blocksize

Block size (MB)	Stale block rate (%)
2	8.33
3	12.67
4	18.78
5	22.25
6	25.97
7	27.17
8	29.88
9	31.46

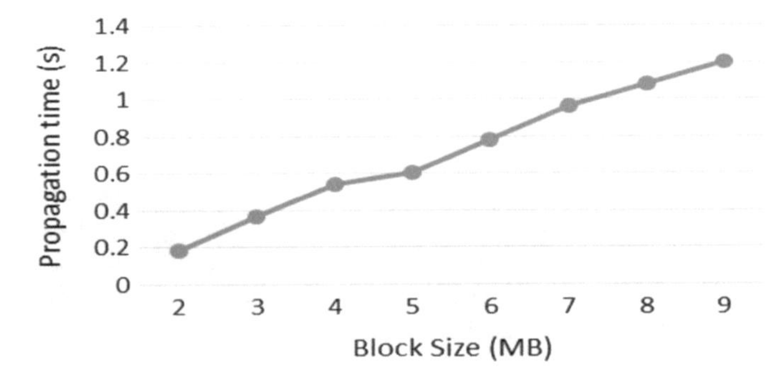

Fig. 2 Average block propagation time with increase in block size

5.1 *Discussion*

A number of solutions have been recommended to mitigate BGP IP prefix hijack attack. They use a centralized database or centralized Public Key Infrastructures as their core mechanism. These solutions are prone to a single point of failure or compromise by cyber attackers. Other solution rely on deployment of a secure version of the BGP routing protocol (SBGP or SoBGP) that requires a complete overhaul or replacement of the BGP routing protocol presently in use on the Internet today. In this chapter, we propose an approach that utilizes a decentralized, immutable database that cannot be compromised. The main motivation for this chapter, is to propose a solution that works with the BGP routing protocol in use on the Internet of Things, without replacing its protocols with a more secure version. The solution proposed in this chapter is capable of mitigating BGP prefix hijack attack initiated by announcing IP BGP sub-prefix, exact prefix or super prefix. It is also capable of mitigating BGP prefix hijack attack initiated by manipulating AS path information contained in a BGP announcement. The stale block rate and propagation time were used to recommend a maximum acceptable block size that is not prone to attack.

6 Conclusion

In this chapter, we proposed a blockchain solution to secure the BGP routing protocol in the Internet of Things. A distributed or decentralized immutable database is created that rely on consensus of participating Autonomous System (AS) to store data in this database. The database validates route origin and AS path information contained in BGP updates received by an AS BGP peer. This solution gives to each AS the capability to detect and mitigate potential IP prefix hijack attack in real time. In addition, the proposed solution eliminates the need for PKI or root of trust like cryptography.

The proposed method in this chapter should serve as the foundation for using the blockchain to secure the Internet of Things infrastructure.

References

1. Rashevskiy, R.B., Shaburov, A.S.: BGP-hijacking attacks: theoretical basis and practical s-cenarios. In: IEEE Conference of Russian Young Researchers in Electrical and Electronic Engineering, pp. 208–212 (2017)
2. Hari, A., Lakshman, T.V.: The internet blockchain: a distributed, tamper-resistant transaction framework for the internet. In: ACM Workshop on Hot Topics in Networks, pp. 204–210 (2016)
3. de la Rocha, A., Arevalillo, G., Papadimitratos, P.: Blockchain-based public key infrastructure for inter-domain secure routing (2017)
4. Wang, N., Wang, B.: A secure inter-domain routing protocol. In: Second International Symposium on Intelligent Information Technology Application (2016)
5. Vidya, K., Rhymend Uthariaraj, V.: Application of logarithmic keying for securing ASPATH in inter-domain routing. In: 2009 First International Conference on Advanced Computing, pp. 86–92 (2009)
6. Butler, K., McDaniel, P., Aiello, W.: Optimizing BGP security by exploiting path security. In: ACM Conference on Computer and Communications Security, pp. 298–310 (2006)
7. Kent, S., Lynn, C., Seo, K.: Secure border gateway protocol (S-BGP). IEEE J. Sel. Areas Commun. **18**, 582–592 (2000)
8. Sermpezis, P., Kotronis, V., Gigis, P., Dimitropoulos, X., Cicalese, D., King, A., Dainotti, A.: ARTEMIS: neutralizing BGP hijacking within a minute. IEEE/ACM Trans. Netw. (TON) 2471–2486 (2018)
9. Yli-Huumo, J., Ko, D., Choi, S., Park, S., Smolander, K.: Where is current research on blockchain technology? A systematic review (2016)
10. Kumari, S., Ruj, S.: SmartDNSPKI: A Blockchain Based DNS and PKI (2017)
11. Leo Arnason, S.: Cryptocurrency and bitcoin: a possible foundation of the future currency, why it has value, what is its history and its future outlook (2015)
12. Bonneau, J., Miller, A., Clark, J., Narayanan, A., Kroll, J.A., Felten, E.W.: SoK: research perspectives and challenges for bitcoin and cryptocurrencies. In: IEEE Symposium on Security and Privacy., pp. 104–121 (2015)
13. Nakamoto, S.: Bitcoin: A Peer-to-Peer Electronic Cash System (2008)
14. Giancaspro, M.: Is a smart contract really a smart idea? Insights from a legal perspective. Comput. Law Secu. Rev. **33**(6), 825–835 (2017)
15. Gervais, A., Karame, G.O., Wust, K., Glykantzis, V., Ritzdorf, H., Capkun, S.: On the security and performance of proof of work blockchains. 3–16 (2016)
16. Ns-3 Network Simulator: https://www.nsnam.org/docs/tutorial/html/introduction.html

Use Cases and Applications

Blockchain and Smart Contract in Future Transactions—Case Studies

O. Bamasag, A. Munshi, H. Alharbi, O. Aldairi, H. Alharbi, H. Altowerky, R. Alshomrani and A. Alharbi

Abstract Blockchain is an innovative technology that has attracted the attention of many industrial sectors with its decentralization and transparency features. It has capabilities that go beyond cryptocurrency and could be harnessed for numerous applications. This chapter aims at shedding lights on use cases that utilize Blockchain and smart contract in real-life applications such as Real Estate contracts, Secure Certificates, and intellectual property protection through digital licensing.

1 Introduction

Blockchain is a recent technology that has gained unprecedented success in the short time that it has existed with many potential applications. The add-only data structure of blockchain in which new data can be appended to the blockchain and old data cannot be deleted/edited makes it an accurate record of transactions between parties. This feature has not only brought about a solution to the ever-existing problem of sharing resources on the go but also has helped to create a completely transparent system enduring that everyone knows where his document is and knows how to access it from any part of the world. This technology has recently transformed the monetary transaction market too by being the backbone behind the digital currency industry which stands out of having no intermediary and thus no loss of money in the middle.

Blockchain supports multi-party authentication and is considered one of the strongest protection technologies. Due to the existence of controls and standards for users, it is difficult to break these security measures and most importantly signing and digital accreditation.

O. Bamasag (✉) · A. Munshi · H. Alharbi · O. Aldairi · H. Alharbi · H. Altowerky · R. Alshomrani · A. Alharbi
Cyber Security Department, University of Jeddah, Jeddah, Saudi Arabia
e-mail: obamasag@uj.edu.sa

A. Munshi
e-mail: ammunshi@uj.edu.sa

M. A. Khan et al. (eds.), *Decentralised Internet of Things*, Studies in Big Data 71,
https://doi.org/10.1007/978-3-030-38677-1_8

One of the enabling features of blockchain is smart contracts. Smart contracts are trustless contracts that autonomously perform an action when a certain condition is met. Thus, resolving the need for third-party validation of transactions and the conveyance of information.

In real estate market, buying a house for example, parties can negotiate and sign a contract within a short time. However, it will take days for the house to be legally transferred to the buyer because the parties have to involve the government department that is responsible for registration of property transfers. With blockchain, the transaction is recorded on the distributed ledger immediately. From the security perspective, real estate market faces issues regarding securing its transactions, i.e. dual sell and fraud on the ownership. These two problems have seriously threatened the market and, specially, its transformation to e-market. Blockchain, through the use of a smart contract, resolved the need for third-party validation of transactions and the conveyance of information. Therefore, changing the ownership of property and transaction validated by parties who fully trust each other can be done without relying on a trusted middleman.

Secure Certificate (academic or professional) storage and verification is another important application of Blockchain. Organizations spend good amount of time in verifying the validity of the candidates' certificates and being free of fraud or manipulation of its contents. In most cases, the certificate verification method is slow as it is done by manually matching the certificate provided with the original document. Using Blockchain technology, certificates are to be stored in records located on Blockchain. Thus, there is a guarantee that the data stored is encrypted, unchangeable, cannot be manipulated and can only be accessed by authorized persons.

Blockchain and smart contracts can also be used to grant digital licenses to protect intellectual property of authors and owners of the original work. A complete record of the works can be maintained with the time and date stamps. Thus, the ownership of the work can be proved, and the intellectual property licenses can be validated.

In this chapter, we present a review and analysis of research work related to Blockchain technology to understand its principles and working mechanism. We also outline some real-life applications that adopted Blockchain technology, namely, real estate, secure certificate, and intellectual property protection.

2 Blockchain Fundamentals

The idea of Blockchain came into the scene in 1991 when a chain of information was used in an electronic ledger account clearly showing that information has not been changed. The first digital application of Blockchain's idea started in 2008 with a paper published by Satoshi Nakamoto (claimed to be not a real name, and the real author remains unknown). Since then, Blockchain has been associated with the cryptocurrency Bitcoin [1]. Bitcoin was presented as a solution to the problem of dual spending, where a string containing operations sealed with the date of HASH was used and can never be changed [2].

Blockchain technology allowed users to be pseudonyms, which means that their names are unknown, but their accounts are documented and recognized, and all transactions can be seen publicly [1]. Blockchain is also slower than the centralized databases because prior to recording the block in the chain, three tasks must be performed: signature verification, machine consensus, block redundant.

2.1 *Blockchain Working Mechanism*

Blockchain technology works based on a distributed ledger that offers a validation mechanism for transactions through a peer-to-peer network. Furthermore, Blockchain works in a decentralized environment that provides transparency and a high level of security without requiring a third party. Each validated transaction is added as a block to an existing chain of transactions. As a result, the transactions within the chain cannot be altered or deleted. This process was named Blockchain because it resembles a pearl necklace. Each pearl bead represents data or information about a particular process. Each pearl that is added to the end contains part of the previous pearl's information. After a pearl is completed, a new pearl is attached to it, as shown in the following Fig. 1 [3].

Specifically, the Blockchain technique is based on the use of secure digital identities. A secure digital identity is generated using private and public key encryption to generate a digital signature for use as a reference point for this identity, allowing controlling ownership and certification [4].

In addition to the use of secure digital identities, Blockchain also involves the accreditation and confirmation of existing digital transactions on distributed network technology. A distributed network is a network of individuals acting as auditors to confirm various things (the process of accreditation) [5].

Blockchain technology enables combining private and public key encryption with a distributed network to create a new type of digital transaction. The complete operation is as follows: After the digital transaction is created, the sender's private key and public key are used for the receiver. This information is stored in a block.

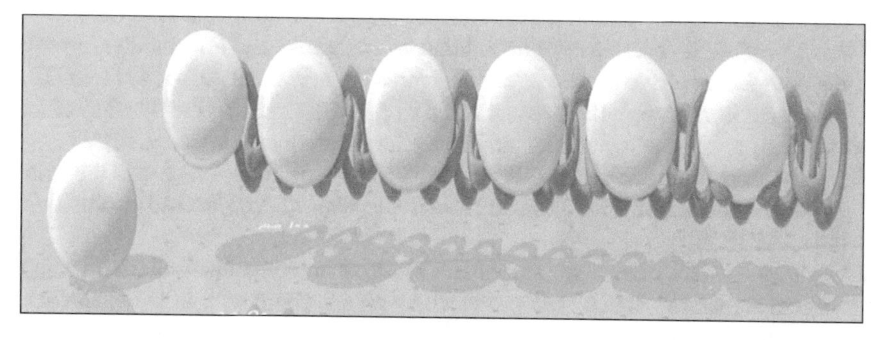

Fig. 1 Blockchain concept

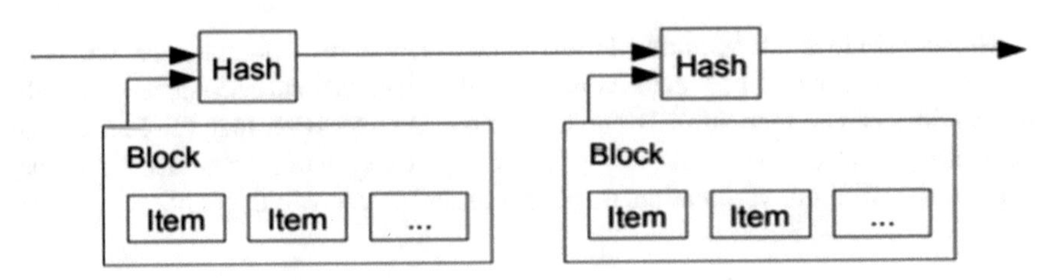

Fig. 2 Structured connection of Blockchain's blocks [2]

This block contains the digital signature and time of this transaction, as well as additional information about the transaction or component parts of the network, as shown in Fig. 2. This part of the network uses transaction auditing. After verifying the transactions and their validity, so as not to repeat transactions or double-spend, the transaction time is stored within the block on the network. Using complex mathematical operations, this block is confirmed or rejected based on whether it contains similar parts to another block. If the block is checked, validated, and found to be unique, it is sequentially added, based on the time stamp, to the blocks that have been previously confirmed to create a blockchain [4].

A blockchain is a decentralized, distributed database that uses cryptography to build a decentralized electronic record of data in a hierarchical, non-editable, manipulative, transparent and easy-to-process format. It also provides the possibility for stakeholders to engage in building, according to the self-administered operating regulations and instructions [6].

2.2 *Blockchain Main Components*

Blockchain consists of four main components: block, information, hash and time. These elements represent the blockchain, and can be explained as follows [7]:

- Block: represents the unit of chain construction, which is a set of operations or tasks to be carried out or implemented within the chain. Examples of blocks are: bitcoin money transfer, registration of data, follow-up situation, or otherwise. Each block will absorb a specific amount of operations and information towards the completion of the processes and then create a new block linked to them. The main goal is to prevent the conduct of fake operations causing the freezing of the chain or prevent the registration and hence termination of transactions.
- Information: is the sub-process that takes place within a single block or is the "Single Order" that is inside the block and represents, with other orders and information, the block itself.
- Hash: is the DNA characteristic of the blockchain, sometimes referred to as a "digital signature". It is a code produced by an algorithm with-in the blockchain program called the "Hash mechanism" and has four main functions:

  - The chain is distinguished from other chains and each series having a distinctive and unique hash.
  - Identify each block and distinguish it from others within the chain, where each block also takes its own hash.
  - Marking information into the block with a unique hash.
  - Linking the blocks together in the chain, where the block is linked to previous and subsequent hashes.

- Time imprint: the time at which any operation takes place within the chain.

The first practical application of this technology is the encrypted virtual currencies, especially the bitcoin. According to the researchers and technical experts, it has not proven any breakthrough in this technology during the ten years of its lifetime since 2008. Although it is still in its infancy and in the development stage, it is considered a major achievement in the information security sector. For these reasons and others, there is a global trend by countries and major institutions to adopt blockchain in many sectors, including finance, financial intermediation, education, health and others [8].

Blockchain has not only been used in digital currency Bitcoin, but also applied in financial and administrative operations linked to at least 19 financial, administrative and service areas. These include banking operations, payments, insurance, logistics and information security, internet of things, health services, energy management, asset management, charitable activities and elections. IBM said that by the end of 2018, at least 15% of the world's banks will be relying on blockchain technology to manage the data of some of their products and services. UBS, Citi Bank, Barclays and Goldman Sachs have already begun to create jobs for engineers responsible for blockchain projects [9].

2.3 *Blockchain Development Phases*

During the past ten years, there have been significant developments in the technique of blockchain will be mentioned and listed in chronological order as follows:

- *2008*: as mentioned above, a person known as "Satoshi Nakamoto" published a study explaining "peer-to-peer" and peer-to-peer systems that enable users to interact directly without the need for an intermediary [2].
- *2009*: the first version of the Bitcoin client program was launched, and a team was created and communicated through a mailing list to develop the project. He was dealing with the team that founded it on the open source of the program did not disclose anything that could reveal his identity [8].
- *2011*: Silk Road website, a black market and the world's first blockbuster market, was launched. It is known as a platform for the sale of illicit drugs, where a "block" coin was used to trade on this platform [2]. At the same time, Douglas Figgelson of Beatles filed a patent application for the "creation and use of digital currency" at the United States Patent and Trade-mark Office [10].

- *2013*: the programmer Vittalik Puterin proposed an electronic system called 'Ethereum', with a digital currency named 'ether'. This online currency was funded and developed by the masses in mid-2014, and the platform began operating on July 30, 2015. Developers have always tried to find a way to differentiate between Ethereum and bitcoin and made the coin exclusive with characteristics that stand out in the market. Ethereum makes it easy to create smart contracts, a self-contained code that developers can use in a wide range of applications [11] (further information on smart contracts will follow).
- *2017*: the use of Smart Contract, a special protocol with a clear objective of executing the negotiation and terms of the contract, allows these transactions to perform credibly without the intervention of any third party. These transactions are traceable and irreversible. Smart contracts contain all information about contract terms and the implementation of all the crisis automatically. They can simplify work in various areas of life, including logistics, administration, and even elections.

The technology is currently being used in prominent fields including: Ownership transfer, Voting, Internet of things, Saving Intellectual ownership, Supply chain management, and many more [12].

3 Blockchain Properties and Security Services

Blockchain technology has several properties including being trustless, decentralized, has distributed nature, immutable, and cryptographic security, detailed in the following [13, 14].

- **Decentralized**: Blockchain system is based on distributed ledger technology combining thousands of nodes in a peer-to-peer network serving as an open electronic ledger. Each computer, or node, has a complete copy of the ledger, so one or two nodes going down will not result in no effect on the data. Blockchain is a decentralized technology as no central authority can control it, making it less corruptible. It effectively eliminates the middle-man as there is no need to engage a third-party to process a transaction. A trust does not have to be placed on a vendor or service provider when a decentralized, immutable ledger can be relied on.
- **Transparency**: Data in blockchain is recorded in every block, hence, it is very difficult to be tampered with. If the data in one block has changed, other blocks containing this data will be changed, and most likely becomes invalid.
- **Immutable**: Each block is connected to all the blocks before and after it. This makes it difficult to tamper with a single record because a hacker would need to change the block containing that record as well as those linked to it to avoid detection. This provides integrity of record and insures that no one manipulates it.
- **Enhanced security**: blockchain allows to store data and uses various cryptographic properties such as digital signature and hashing to ensure it's protection. As soon as the data enters a block in a blockchain, it cannot be tampered with.

- **Consensus mechanisms**: Blockchain is fault tolerance as its networks are built in such a way that it assumes any individual node could attack at any time. Consensus protocols, like proof-of-work, ensures that even if the network is attacked, it will complete its functions as intended regardless of human misconduct or intervention.
- **Enhanced security**: Records on a blockchain are secured through cryptography. Participants have their own public/private keys corresponding to the transactions they take part into facilitate a personal digital signature. Because of its distributed nature, file signatures can be checked across all the ledgers on all the nodes in the network and verified that they haven't been changed. If a record has been changed, then the signature is rendered invalid and all peer network are immediately notified. This allows to use the blockchain ledger to verify that data backed up and stored in the cloud with third-party vendors has gone completely unchanged even weeks, months, or years later. Thus, we could conclude that blockchain offers reliable, independent data verification.
- **Resistance to Hackers**: Hackers would have an extremely hard job trying to hack a blockchains. They are decentralized and distributed across peer-to-peer networks that are continually updated and kept in sync. As they are not contained in a central location, blockchains do not have a single point of failure and cannot be changed from a single computer. It would require massive amounts of computing power to access every instance (or at least a 51% majority) of a certain blockchain and change them all at the same time.

4 Smart Contracts

In this section, we will highlight the main concepts of smart contracts, its architecture, properties and operations.

4.1 Definition

The term "smart contract" was first envisioned in1994 by Szabo [15], who defined a smart contract as "a set of promises, specified in digital form, including protocols within which the parties perform on these promises." Szabo, in his known example, analogized smart contracts to vending machines: it takes in coins, and via a simple mechanism (e.g. finite automata), dispense the product and change according to the displayed price. Smart contracts go far beyond this example by embedding contracts in many sorts of transactions by digital means. Szabo also predicted that through clear logic, verification and enforcement of cryptographic protocols, smart contracts could be far more efficient than their paper-based counterparts. However, the idea of smart contracts has not been realized until the emergence of blockchain technology as the public and append-only distributed ledger technology (DLT).

The consensus mechanism makes it possible to implement smart contract in its true sense. In general terms, smart contracts can be defined as the computer-based protocols that electronically facilitate, verify, and enforce the contracts established between two or more parties on blockchain [16]. As smart contracts are typically implemented on and secured by blockchain, they have some inherited characteristics. First, the code of a smart contract will be recorded and verified on blockchain, therefore, making the contract secure against tampering. Second, the smart contract is executed among anonymous, trustless individual nodes with no centralized or third-party authorities. Third, a smart contract acts like an intelligent agent and might carry its own cryptocurrencies or other digital assets, and release them when preset conditions are triggered [17].

Ethereum [6] is the first public blockchain platform that supports smart contracts with the assistance of Turing-complete virtual machine called Ethereum virtual machine (EVM). EVM is the runtime environment for smart contracts, where every node in the Ethereum network runs an EVM implementation and executes the same instructions. Many high-level programming languages, such as Solidity [6], can be used to write Ethereum smart contracts, and the contract code is compiled down to EVM bytecode and deployed on the blockchain for execution. Although smart contracts have progressed greatly in recent years, it still has many security and privacy issues. A well-known attack targeted DAO (Decentralized Autonomous Organization) in June 2016. DAO is an entity operating through smart contracts. It is meant to be a venture capital fund for the crypto and decentralized space. Its financial transactions and rules are embedded in, and secured by Ethereum Blockchain, eliminating the need for a central governing authority. In June 17, 2016, a hacker exploited a bug in the coding that allowed him to drain funds from the DAO. In the first few hours of the attack, 3.6 million ETH (Ethereum's currency) were stolen, the equivalent of $70 million at that time. Once the hacker had done the damage he intended, he withdrew the attack. In this exploit, the attacker was able to "ask" the smart contract (DAO) to give the Ether back multiple times before the smart contract could update its balance [16].

4.2 Smart Contract Architecture and Operations

The operational mechanism of smart contracts is shown in Fig. 3.

Smart contracts are characterized by two main attributes: (1) value and (2) state. The activating conditions and the corresponding actions of the contract terms are defined using provoking condition statements such as "If-Then" statements. Smart contracts are approved and signed by all parties involved and sent as transactions to the blockchain network. Transactions are then broadcasted via the network, verified by the blockchain miners and saved in the designated block of the blockchain.

The generators of the contracts get the resulting parameters (e.g., contract address), then users can trigger a contract by sending a transaction. System's incentive mechanism encourages miners and will share their computing resources to validate the

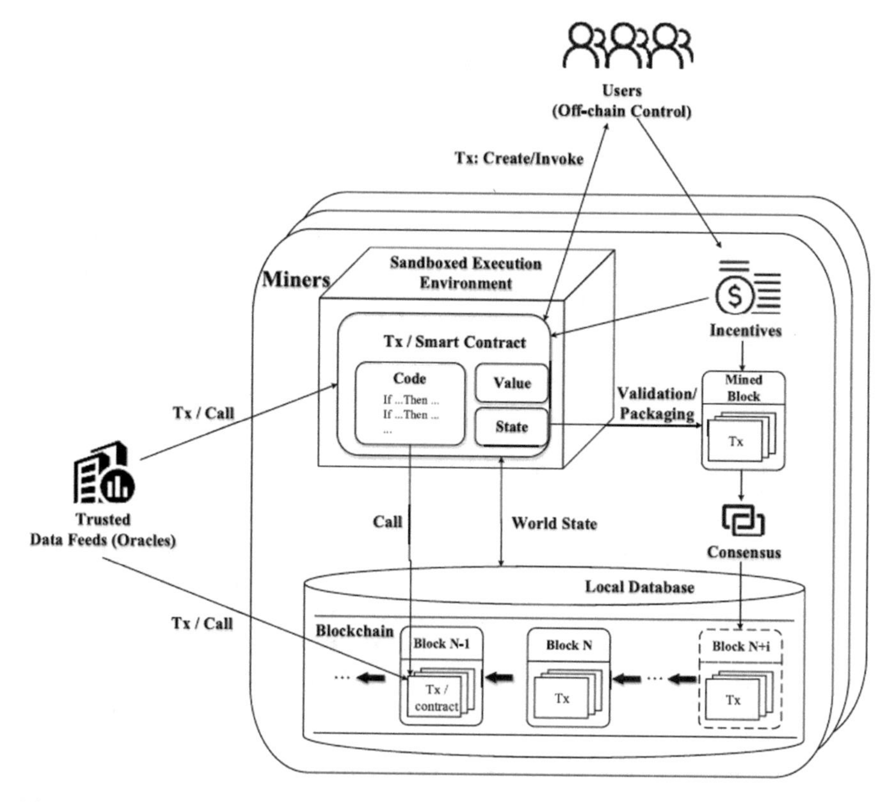

Fig. 3 Operational mechanism of smart contract [16]

transaction. That is, after the miners receive the contract generation or invoking transaction, they generate contract or run contract code in their local Sandboxed Execution Environment [(SEE), e.g., EVM]. According to the input of trusted data feeds, i.e. Oracles, and the system state, the contract decides whether the current scenario meets the invoking conditions. If the conditions are met, the corresponding actions are executed. Upon a transaction validation, it is stored into a new block. The new block is chained into the blockchain once the whole network agrees on a consensus.

In the following, Ethereum is presented as an example to introduce the operational process of smart contracts. Ethereum is currently considered the most widely used smart contracts development platform that can be seen as a transaction-based state machine: it begins with original states and executes transactions incrementally to transform it into some final states. The goal is for the final states to be accepted as the approved "version" in Ethereum environment [18]. Ethereum introduces the concept of accounts and defined two types of them: (1) externally owned accounts (EOAs) and (2) contract accounts. The difference is that the former is controlled by private keys without code associated with them, while the latter is controlled by

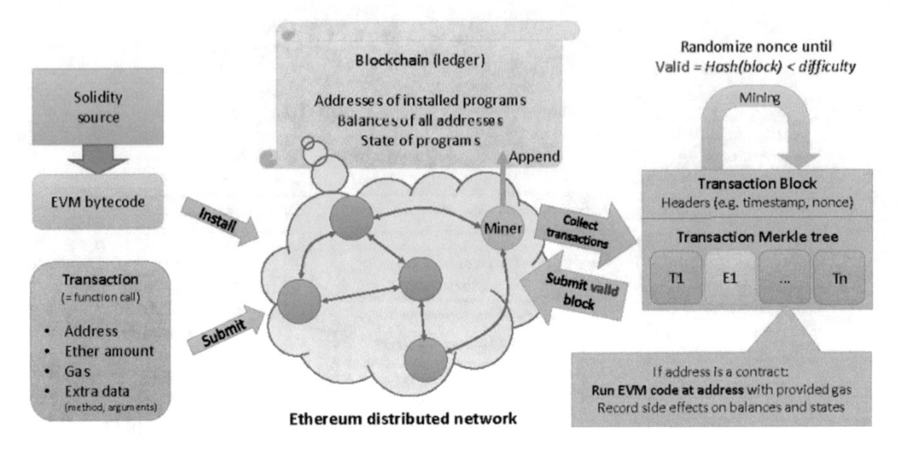

Fig. 4 Smart contract workflow in Ethereum Network [19]

their contract code with associated code. Users can initiate a transaction only via an EOA. The transaction can include binary data (payload) and Ether. If the recipient of a transaction is the zero-account Ø, a smart contract is created. Alternatively, if the receiver is a contract account, the account will be triggered and its associated code is executed in the local EVM (the payload is provided as input data) [19]. The transaction is then broadcast to the blockchain network where miners will verify it, as shown in Fig. 4.

To refrain from issues of network abuse and to avoid the inevitable problems stemming from Turing completeness, all coded computations (e.g., generating contracts, initiating message calls, managing account storage, and running operations in the virtual machine) in Ethereum is subject to fees, which is considered as a reward for miners who provide their computing resources. The unit used to measure the fees required for the computations is called gas [18].

5 Applications of Smart Contracts

5.1 Real Estates

Despite the fact that real estate market is flourishing, it faces security issues like dual sell and fraud on the ownership. These two problems are very serious and need to be addressed for the success of this key market. Blockchain technology has provided solutions to these issues due to its transparency and reliability in the peer- to- peer network. Smart contracts, as explained earlier, eliminated the need for third-party validation of transactions and the conveyance of information. Therefore, changing the ownership of property and transaction validated by parties who fully trust each other without relying on a trusted middleman [20].

Researches [20, 21] have employed blockchain technology in improving the land registration systems. There are two types of land registration system: Recording and Title systems, both have seen improvement in their implementation and performance. One of the issues that Blockchain has contributed in solving is the problem of "double spending" of the recording system, which is done by applying the sequence transaction [20].

5.1.1 System Overview

The proposed real estate system, named 'My Land', is implemented through a web application employing blockchain, where cryptographic mechanisms are applied to real estate transactions, shown in Fig. 5. Blockchain provides proof of the property, which is considered a critical aspect in the real estate market. In addition, Blockchain improves the level of transparency, and almost eliminate fraudulent activities. It also speeds up land registration, by a more straightforward process, and provide for an open approach to data access [20].

Smart contract is the ideal framework for our system as it automatically execute when certain conditions between the two parties are met. There is no need for trusted third-party (such as lawyer, broker), since these contracts meant to be self-executed. It

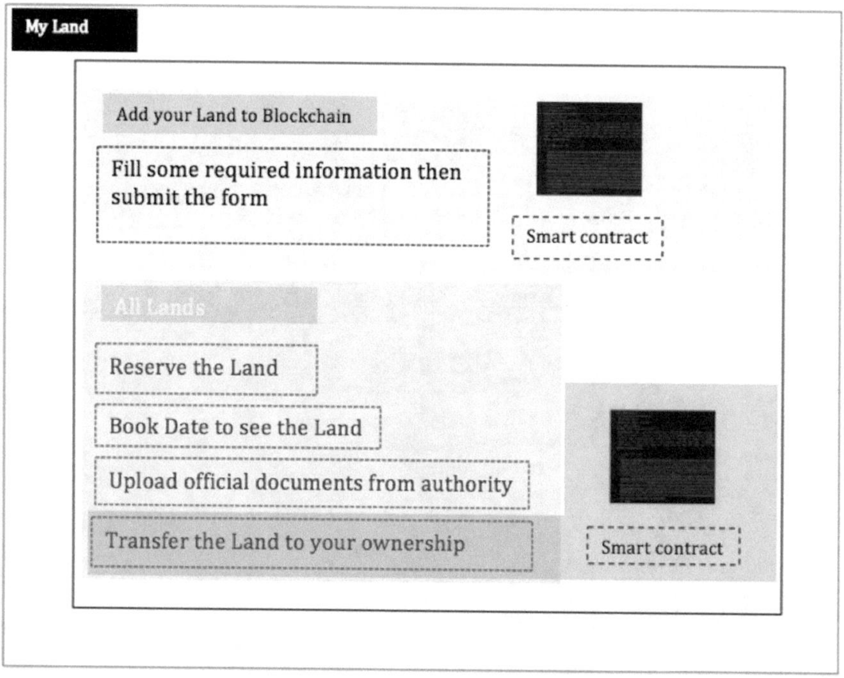

Fig. 5 My Land system framework

will be employed to make the land transfer process effective, verifiable, and publicly visible to all users of the system. The seller and the buyer register on the site, where the seller offers his property through filling an electronic form built within the system. The buyer purchases the land by first booking it and then contact the seller to inspect the land and complete the transactions.

Figure 6 explains the flow of data in the system. The data here means ownership of the land, transferring from the seller to the buyer after passing through several processes aimed at verification before the transfer of ownership. Note that the first action done in the system is registration of users, after the registration phase, the processes in the system distributed into several parts. **First**, related to the seller such as land registration and proof of ownership of the land, and then offer the proposed price. The **second** part of the process is for the buyer to register the general information, then booking the land if desired to buy. After following the necessary governmental procedures to register the contract, **third** part is launching the smart contract between the parties involved. If the conditions are fulfilled, such as the presentation of an instrument on behalf of the buyer and the delivery of the payment and the approval of the seller, system will lead to final step. **Finally**, the land ownership is transferred to the buyer, and new information added to the blockchain.

One of the most important tasks conducted by the system is the transfer of ownership of the land. It is carried out through a number of steps executed by both the seller and the buyer to be transferred, as shown in Fig. 7.

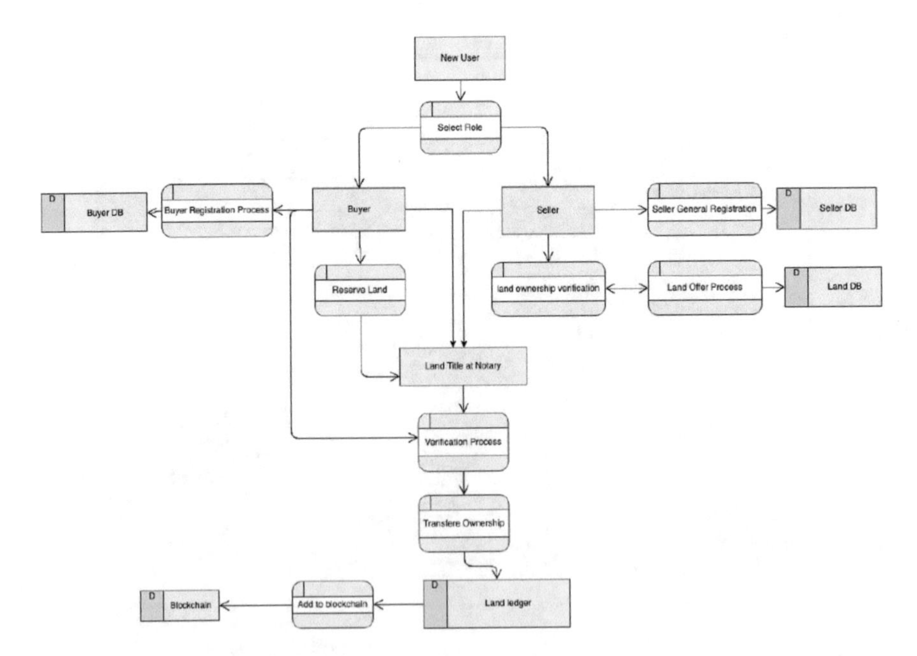

Fig. 6 Data flow diagram of 'My Land' system

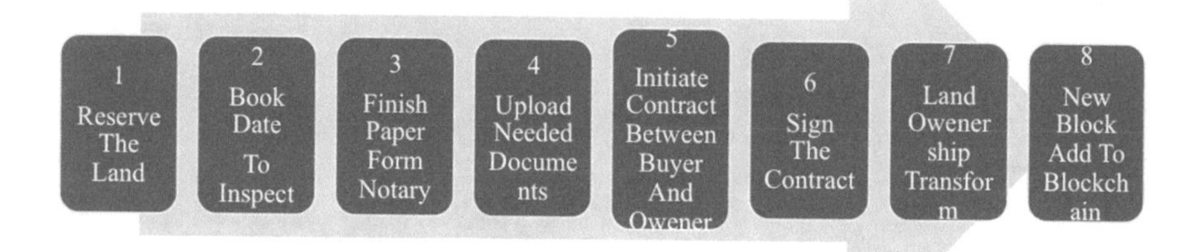

Fig. 7 Ownership transfer process

5.1.2 System Implementation

This subsection will provide an overview of the implementation of web application for real estate, named 'My Land'. It will also outline the tools used to accomplish the intended functionality of the system, i.e. buying, selling and saving the ownership of property within a trusted and secure platform.
Tools:

- *Ethereum Platform*: is a decentralized platform that supports Smart Contracts: where the ownership of property is transferred easily and protected from any fraud or third-party interaction. Our project runs on Blockchain, which is a shared infrastructure that can be used to store the ownership of property.
- JavaScript: It helps establishing the front-end of My Land system, which support the use of (Web3 API) and allow web application to interact directly with the smart contracts. One of the important features of JavaScript is Web 3.js, which is a class containing all the models related to Ethereum.
- MetaMask: is an environment that allows running an Ethereum distributed application DApps in any browser without the need for downloading a full Ethereum node. MetaMask provides a user interface to manage the user's identities and track user's transaction. Ropsten Test Network have been used to test the application before deploying it in Ethereum.
- Solidity: is a high-level programming language for creating smart contracts with a syntax similar to JavaScript. Ethereum supports Solidity, hence, it could be handled via JavaScript files.
- Remix: is a browser for smart contracts based on Solidity language. It runs within the normal browsers, compiles smart contracts, and help to connect JavaScript to the contract address in the Blockchain. JavaScript must include smart contract API in order to execute its functions.

The main part of our project is creating a smart contract. We use Solidity language to write our contract and we used Remix However, the contract on the Ethereum Blockchain must be mined that required some gas, which is a reward for the Miner.

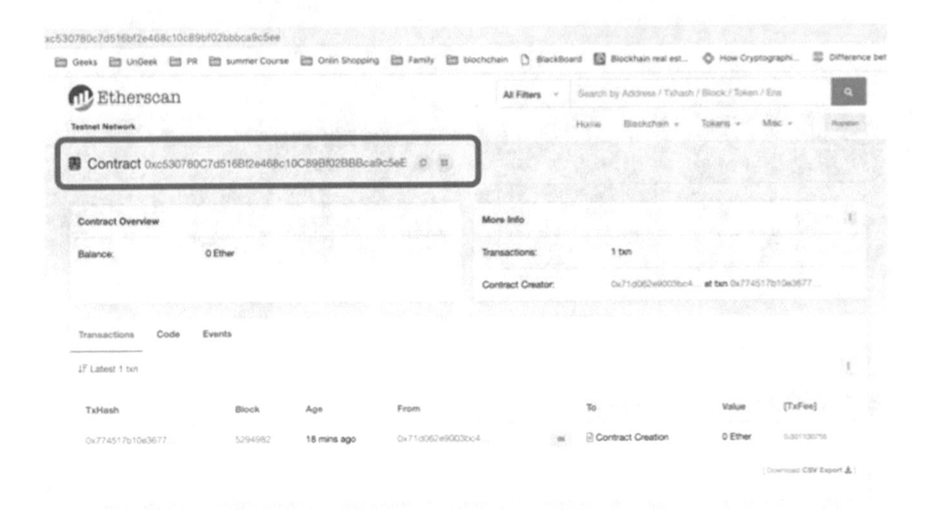

Fig. 8 Transaction on smart contract address

Creating Smart Contract:
Smart contract in My Land system was created to transfer the property from a user's address to another address within the network of Ethereum. We took into consideration that the user may have more than one land. As mentioned earlier, we are using Ropsten test network in order to test our smart contract before deploying on the main network of Ethereum. To do this, we need an address for the smart contract, which can receive data or money in the future. An address for our smart contract can be obtained by compiling the smart contract in Remix and then run the contract as a result. Figure 8 shows the obtained address of the contract, which enable performing a transaction.
Creating a New Account on Metamask:
Each Ethereum account has an address to enable accounts of smart contracts or others to communicate among them. Each user will receive a secret key that has to be saved securely. The address of each account is unique and never duplicated, see Figs. 9 and 10.

5.1.3 Walk Through 'My Land' System

This section will present the flow of our web application from user perspective and the description of each page.

The home page of My Land website provides users with the main services such as login, land index and chatting (Fig. 11). Therefore, they will be able connect easily with Ethereum Blockchain and experience transferring the ownership of property securely. It is worth mentioning that the ownership of the land will not change unless some conditions are met, i.e. providing proof of payment such as check. As a result,

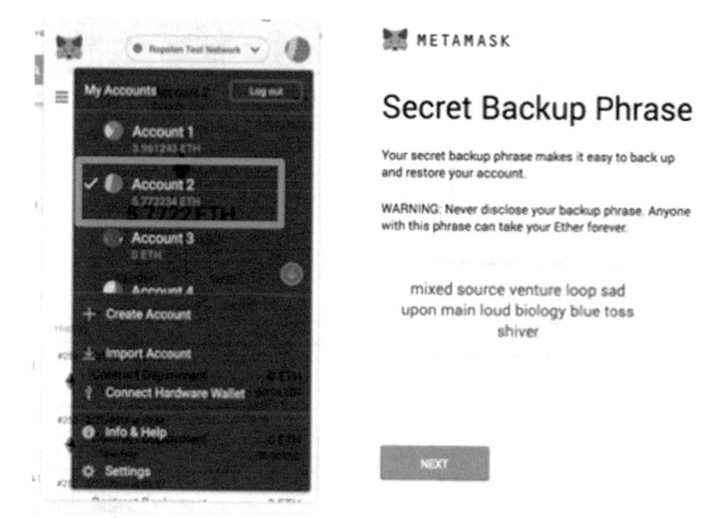

Fig. 9 MetaMask account and secret phrase

Fig. 10 The address on MetaMask account

the admin will allow the user to change the ownership to his/her name by adding the property to Blockchain.

In order to utilize the web site services, users must sign up and fill some required information such as First name, Last name, National ID, E-mail, and Password, as shown in Fig. 12.

Fig. 11 My Land web application home page

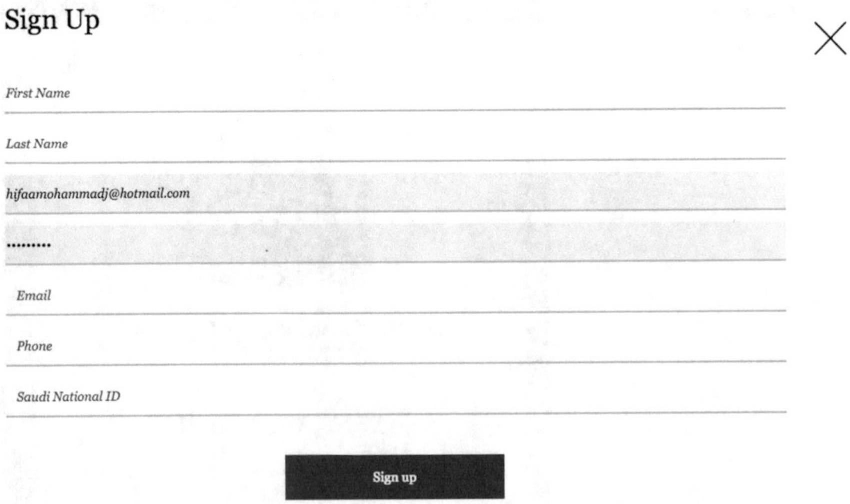

Sign Up

First Name

Last Name

hifaamohammadj@hotmail.com

•••••••••

Email

Phone

Saudi National ID

Sign up

Fig. 12 Signup form

Then the user has to login by the E-mail and password he/she entered while registration phase. See Fig. 13.

After singing up, the user is allowed to browse our web page. There is an option for the user to add his ownership to the Ethereum blockchain by entering some required information such as land Saudi National address, land area, land price. Also, upload images for the land and then submitting the form, as in Fig. 14.

Fig. 13 Login form

Fig. 14 Adding land to Blockchain

"Add Your Land to Blockchain" form creates an address for the land on Ethereum blockchain. So, it simplifies the process of interaction with our smart contract, which is deployed by Remix IDE where all lands are added to database and automatically connected by land address. Figure 15 shows the page of "Find Land" which presents the information of lands that are on sale.

User selects the desired land to purchase by clicking the button "Buy Now". Then the user will have to follow five steps to complete transferring land's ownership and stores it in Ethereum Blockchain. Our web application is designed to connect the user directly with his/her account of Ethereum Blockchain where the address of the user is presented as if he/she is already signed in MetaMask. Additionally, there is an option to install MeataMask if the user does not have an account, as in Fig. 16.

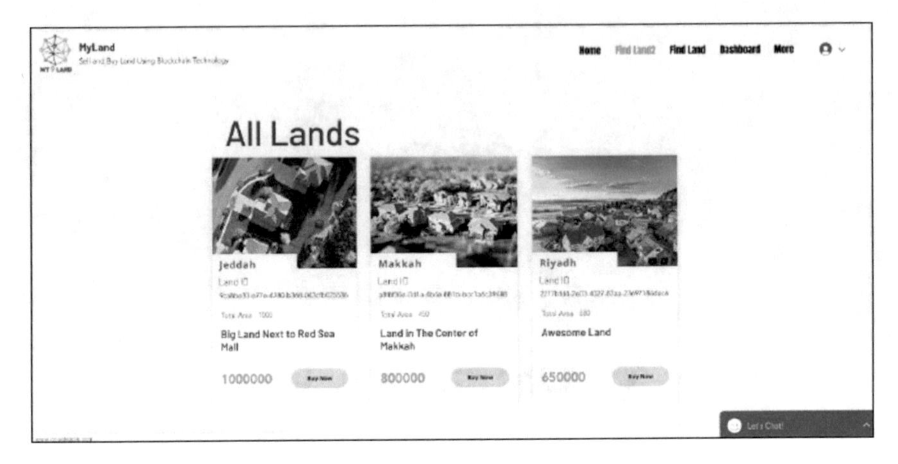

Fig. 15 Find land page

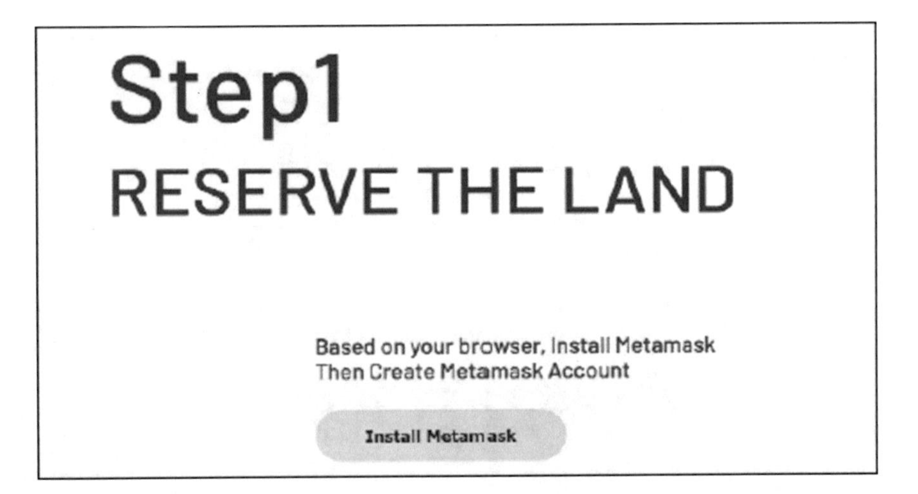

Fig. 16 Step 1: Reserve the land

The second step (Fig. 17) is to select a date for meeting with seller and viewing the land.

The user (buyer) will have to upload the deed and the receipt, which prove the ownership of the property, as shown in Fig. 18.

Then the user will move to the fourth step which allows him to transfer the property to his ownership by entering Land ID as indicated in Fig. 19.

Within this step the user will experience the Blockchain technology by pressing the bottom of "Transfer Ownership". As a result, MetaMask notification window will pop up to the user in order to confirm the transaction. For the user to confirm the transaction, he/she must be sure to have enough Ethers to complete this transaction,

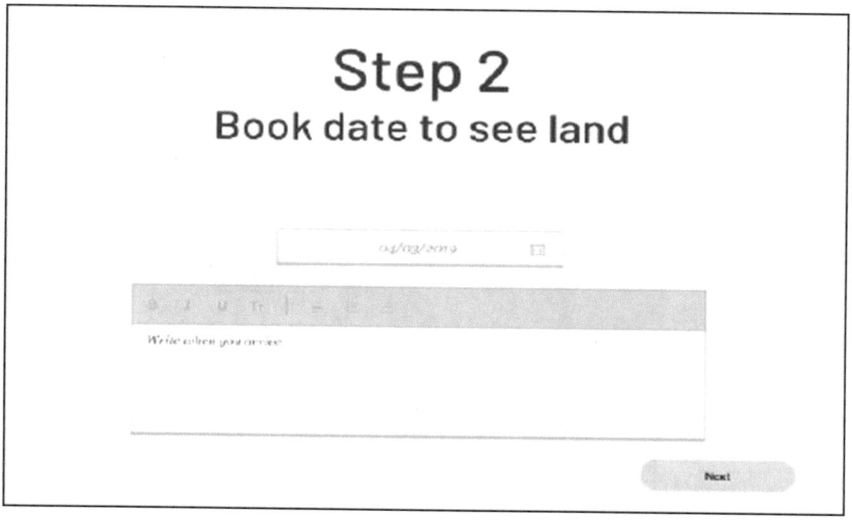

Fig. 17 Land viewing appointment

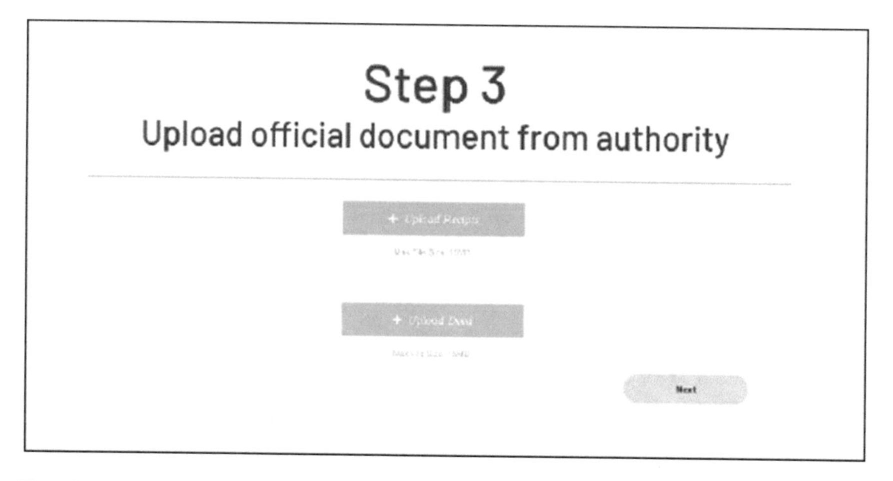

Fig. 18 Step 3: Documents upload

in terms of gases (a gas is special unit used in Ethereum). As shown in Fig. 20, the land will transfer to Account 2 using our smart contract.

According to Ethereum Blockchain, after the conformation is done, the user is able to see the progress of his transaction by using the EtherScan.com, as shown in Fig. 21.

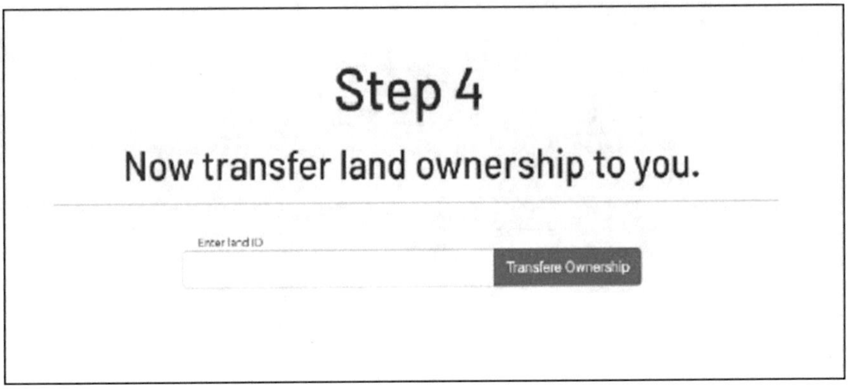

Fig. 19 Step 4: Transfer of land ownership

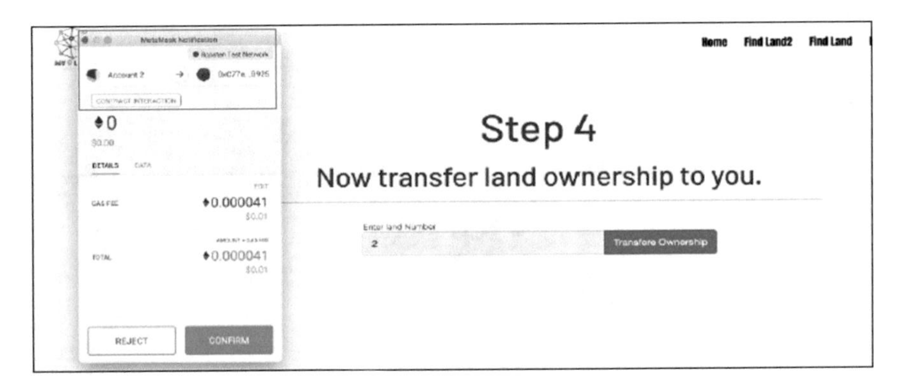

Fig. 20 Metamask notification

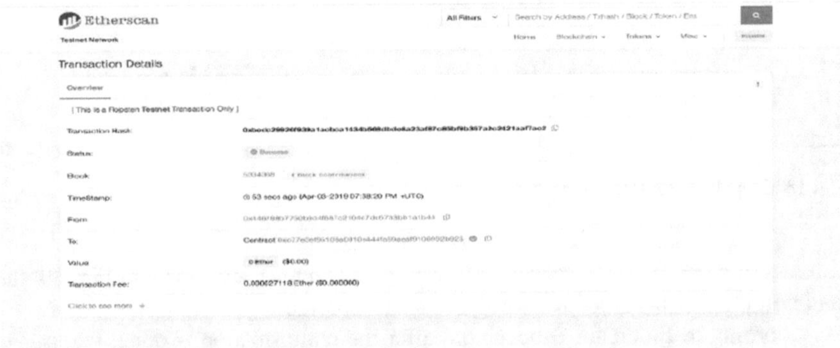

Fig. 21 Track transaction with Etherscan

5.2 Certification with Blockchain

Certificates perform an essential role in both educational and commercial sectors as it can attest for its owners' qualifications and skills [22]. Certificate validation could be an issue that a person may face while applying for a job at a company. Although this is quite a straightforward process and should not take much of a time with legitimate certificates, the complex and long steps to verify the certificate by these companies becomes very annoying for the applicants [23].

As verification of certificate's validity is based on integrity and authentication of origin, Blockchain can be utilized in such an application. It supports multi party authentication and data integrity and considered one of the strongest protection technologies [24]. In this section, we will present BlockCert UJ, a certificate generation and validation application, for the University of Jeddah using Blockchain technology. The idea of our project BlockCer UJ, is inspired from an open source project called BlockCert [25]. BlockCert UJ transforms the classical paper-based certificates into digital ones protected by Blockchain.

5.2.1 System Overview

Following are the main objectives of the system.

- Store students' certificates in a record on Blockchain.
- Each students' certificate is encrypted on Blockchain.
- Students can easily share their documents with people or institutions.
- Companies and other institutions can easily verify the certificate of an applicant.

Figure 22 shows the main components of the system and their interactions.

5.2.2 Implementation

- Python language: Python is an interpreted, object-oriented programming language. Python is easy to learn and relatively portable. In addition, the source code is freely available and open for modification and reuse.
- JavaScript: JavaScript is a programming language developed to display dynamic websites. It relies heavily on Java.
- Platform, we work on ubuntu as it flexible and good environment for the developer.
- Blockchain technology: Blockcerts was built as a technical standard to work across any blockchain, Bitcoin blockchain and Ethereum.
- Web application we use Dejango.

The blockchain certificates is issued by creating a transaction from the issuing institution to the recipient on the Ethereum blockchain that includes the hash of the certificate itself.

The home screen consists of three users, must choose one of them:

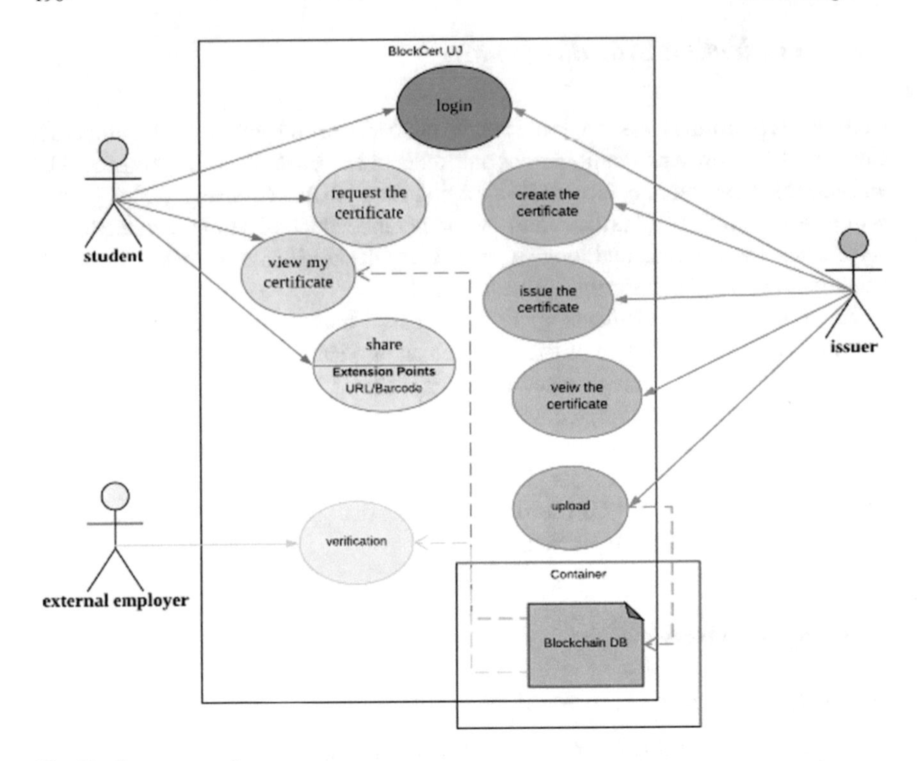

Fig. 22 System overview

- The owner of the certificate: On the Owner page, a login will be available for the user who is already registered (Fig. 23). The owner's home page has two options to perform 'request the certificate' and 'viewing the certificate'. Once the certificate is issued, a notification massage is sent to the owner's e-mail. However, owners are not allowed to request the certificate unless they have an Ethereum address. So, for simplicity we provide a direct link to create Ethereum address. After obtaining the public address, owner is asked to fill the rest of the data then add the information to the requested list on the employee page.
- The issuer of the certificate: following steps are performed by a certificate issuer:
  - Sign in: If a new user option is selected, a page opens to fill the data and then goes to the certificate issuer home page.
  - Log in: If you choose the log in option, enter the correct user name and password and go to the certificate issuer home page.
  - View certificates: Recently issued certificates are displayed on this page for employee.
  - create certificate: the employee must enter the required students' information then the certificate is ready for the issuance so, at this stage the certificate will not be signed.

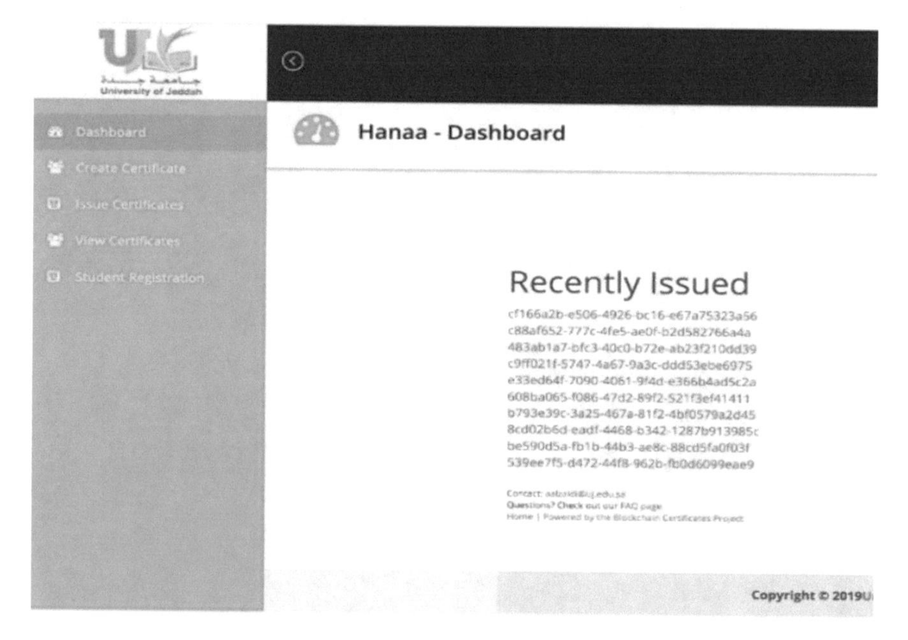

Fig. 23 Owner's home page

- From the button of issuing Certificate we can issue the certificate and upload it in Blockchain, as shown in Fig. 24.
- When clicking on the blockchain address, the transaction details will be displayed as represented in Fig. 25.

Verify the certificate: The page contains a specific field to insert a URL or JSON file of the certificate. After clicking on the upload button, the user will be able to view the certificate. As shown in Fig. 26.

5.3 Digital Licenses for Protect Intellectual Property

Blockchain technology has enormous potential for managing intellectual property rights. Registering intellectual property rights in a distributed registry rather than a traditional database would translate them into "smart intellectual property rights".

Smart contracts can be used to prepare and implement intellectual property agreements, such as licenses, and allow the timely transfer of payments to rights holders; intellectual property "smart information" on protected works, such as songs or images, can be encoded in digital format or image).

Creative Commons [26] is a non-profit organization based in San Francisco, USA, which aims to expand the scope of creative work available to people to exploit and build on it in compliance with the requirements of intellectual property laws, and

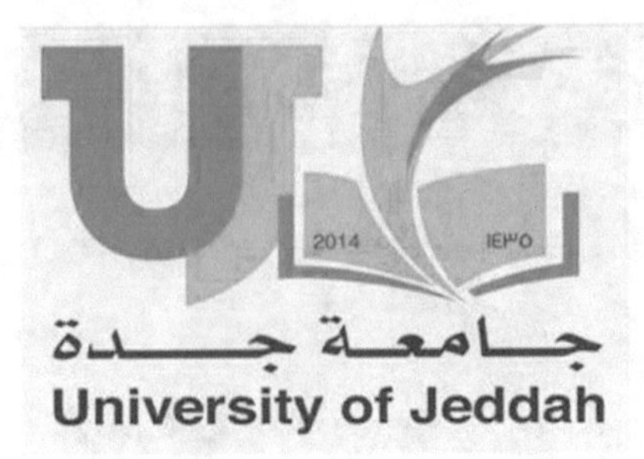

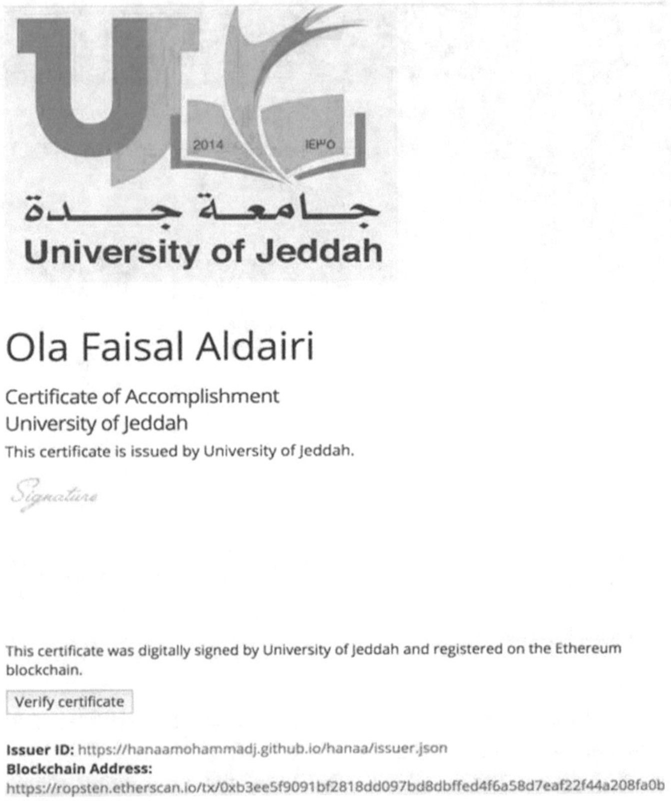

2014
جـــامـعـة جـــدة
University of Jeddah
Ola Faisal Aldairi
Certificate of Accomplishment
University of Jeddah
This certificate is issued by University of Jeddah.
Signature
This certificate was digitally signed by University of Jeddah and registered on the Ethereum blockchain.
Verify certificate
Issuer ID: https://hanaamohammadj.github.io/hanaa/issuer.json
Blockchain Address:
https://ropsten.etherscan.io/tx/0xb3ee5f9091bf2818dd097bd8dbffed4f6a58d7eaf22f44a208fa0b

Fig. 24 Student certificate

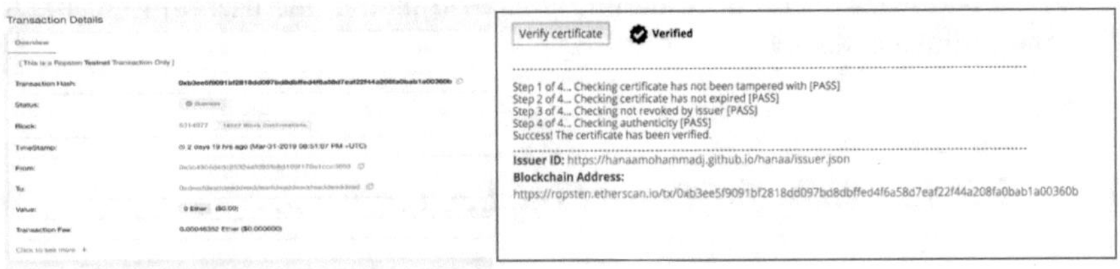

Transaction Details
Verify certificate
Verified
Step 1 of 4... Checking certificate has not been tampered with [PASS]
Step 2 of 4... Checking certificate has not expired [PASS]
Step 3 of 4... Checking not revoked by issuer [PASS]
Step 4 of 4... Checking authenticity [PASS]
Success! The certificate has been verified.
Issuer ID: https://hanaamohammadj.github.io/hanaa/issuer.json
Blockchain Address:
https://ropsten.etherscan.io/tx/0xb3ee5f9091bf2818dd097bd8dbffed4f6a58d7eaf22f44a208fa0bab1a00360b

Fig. 25 Transaction details

issue 6 types of licenses through the website, supports all languages and uses the method For traditional Internet applications, the license is not associated with the name of the author and does not guarantee any legal rights.

Licenses of Creative Commons is an image placed on the work to be licensed, without linking any information about the license holder. Creative Commons does not give any other services such as resource conservation or tracking or even a control.

Fig. 26 Certificate verification

This has become a source of concern for publishers and authors where their works can be robbed, altered and not recognized.

The system proposed here, named 'CCBlockchain', aims at employing blockchain to provide secure and efficient intellectual property licensing through the following objectives.

Keeping authors' records on the Blockchain, each author will be assigned a unique records, each time the work is modified it will be saved with a new value. When keeping the work register on blockchain, no other person can assign the work to himself since he will return to the original author.

The network of nodes on the Blockchain through which the recovery of any work through the keys to encryption and can be shared and therefore will not be lost any work, which will achieve the sustainability of the work. Hash value can be added hidden in each work raised so that the work can be tracked on any existing site and cannot duplicate the work or affiliation to the original author.

Implementing digital licenses using the blockchain technique is realized using distributed records to create "intellectual property IP records", through issuing non-changeable, high-security digital licenses to original authors and owners that include the most accurate details of ownership and work. The application can also verify the ownership of the works and the validity of the intellectual property licenses.

The public network of blockchain (Ethereum) and smart nodes were employed to implement the application of protecting intellectual properties. The application is public and available to all interested researchers, authors and others to maintain intellectual property rights for their works.

One of the most popular applications launched in this field is the launch of "We-mark" based on the blockchain technology to manage image rights and its own encrypted currency [27]. Copyrightbank [28] is a web application recording the intellectual property rights to the block of all file types. It supports English and a special network type and for a fee.

5.3.1 System Overview

The application consists of the following main components (shown in Fig. 27):

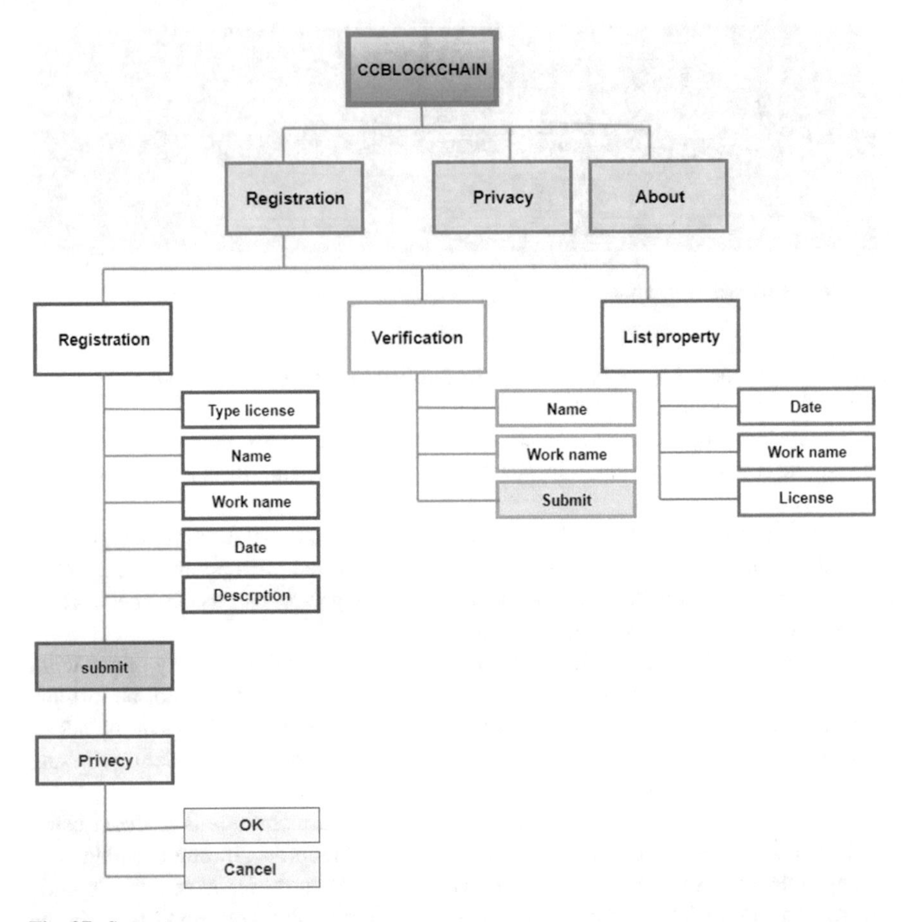

Fig. 27 System structure

- Registration: The system issues multiple licenses as per the user's choice, where these licenses are reserve the ownership of his work.
- Verification of ownership: User can verify the validity and validity of the licenses.
- Privacy: User can learn about license policies and site policies.
- List Properties: user can review his registered works and their licenses.

5.3.2 System Implementation

In this section, we will explain the implementation process of CCBlockchain application by listing the employed tools and a walkthrough the implemented system. Following are the tools and programming languages used to implement the system.

- HTML, Cascading Style Sheets (CSS),

Fig. 28 CCBlockchain homepage

- jQuery, Java script
- Web3
- Solidity
- Metamask
- Ethereum Platform
- Remix IDE (for create smart contract)
- Visual Studio Code
- Godaddy (for hosting application).

CCBlockchain homepage (Fig. 28) consists of:

- Introductory video clip
- Registration of intellectual property rights to work

This page informs the user through the video of the services provided by the website. After the user select Get Start, the main menu will appear containing the following options:

- Registration
- Verification
- List of Property

When the user selects to register, the data to be entered will appear to record the work data and personal data in the blockchain, see Fig. 29. After filling in the work data and personal data and pressing the submit button, the user will see a pop-up window that reminds him of the privacy data and upon approval the data will be recorded.

After the data is recorded in the blockchain the work license will appear containing all the data together as Hash value (see Fig. 30).

In case user select verify button to validity of the license will show him the data to be entered (see Fig. 31).

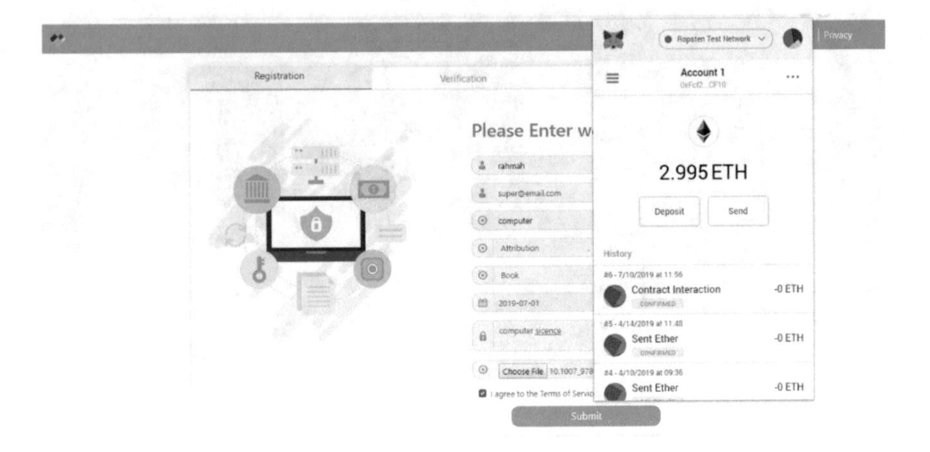

Fig. 29 Registration page

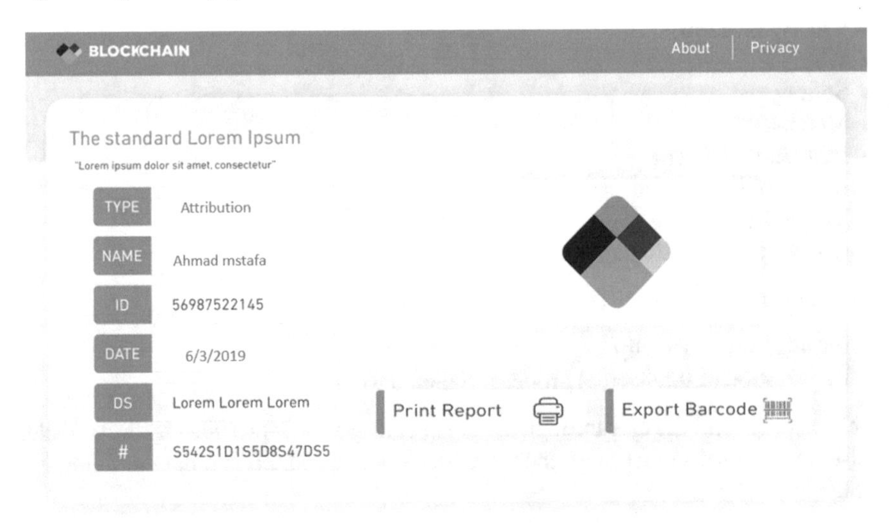

Fig. 30 Blockchain work license

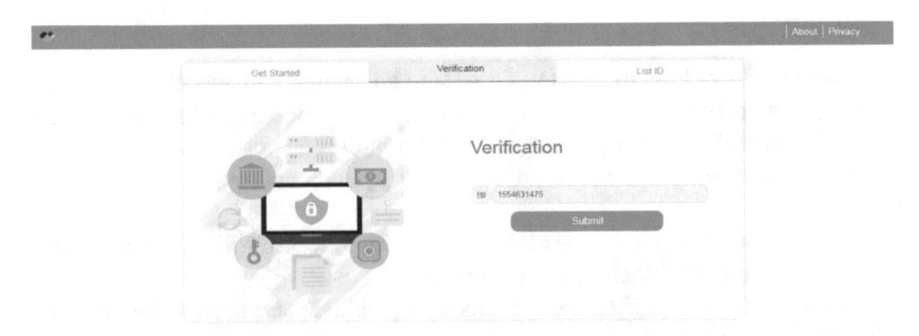

Fig. 31 Verification page

BLOCKCHAIN About | Privacy

REGISTRATION VERIFICATION LIST ID

LIST ID

DATE	NAME	LINK
14/02/2019	Mohamed Ahmed	https://www.google.ps
14/02/2019	Mohamed Ahmed	https://www.google.ps
14/02/2019	Mohamed Ahmed	https://www.google.ps

Fig. 32 List property page

If the user wishes to view the history of the property he can do so after selecting the property list (Fig. 32).

6 Conclusion

This chapter presented the principals and techniques behind Blockchain technology, with an emphasis on Smart Contract. The chapter has also explored applications of blockchain and smart contract to advance the IT industry, namely Real Estate, Verifiable Certificates, Intellectual Property Protection.

Blockchain and smart contract will continue to penetrate in various types of industry and will be one of the main contributors to the international economy.

References

1. Yaga, D., Mell, P., Roby, N., Scarfone, K.: Blockchain Technology Overview. National Institute of Standards and Technology, Gaithersburg, MD (2018)
2. Nakamoto, S.: Bitcoin: A Peer-to-Peer Electronic Cash System. https://bitcoin.org/bitcoin.pdf (2008)
3. Rennock, M., Cohn, A., Butcher, J.: Blockchain technology and regulatory investigations. The Journal **1**, 35–45 (2018)
4. Khudnev, E.: Blockchain: Foundational Technology to Change the World, Bachelor's Thesis, School of Business and Culture, Lapland University of Applied Science. https://www.theseus.fi/bitstream/handle/10024/138043/Evgenii_Khudnev_Thesis.pdf?sequence=1&isAllowed=y (2017)

5. Grech, A., Camilleri, A.: Blockchain in education. JRC, Luxembourg (2017)
6. Bruyn, S.: Blockchain an introduction. University Amsterdam, Amsterdam (2017)
7. Melanie, S.: Blockchain: Blueprint for a new economy, 1st edn. O'Reilly Media (2015)
8. Xethalis, G.E., Moriarty, K., Claassen, R.: An Introduction to Bitcoin and Blockchain Technology. Kaye Scholer, New York (2018)
9. Mencias, A., Dillenberger, D., Novotny, P., Toth, F., Morris, T., Paprotski, V., Dayka, J., Visegrady, T., O'Farrell, B., Lang, J., Carbarnes, E.: An optimized blockchain solution for the IBM z14. IBM J. Res. Dev. **62**, 1–4 (2018)
10. Venter, H.: Digital currency—A case for standard setting activity, Australian government (2016)
11. Vujičić, D., Jagodić, D., Ranđić, S.: Blockchain technology, bitcoin, and Ethereum: A brief overview. In: International Symposium INFOTEH-JAHORINA (2018)
12. Karamitsos, I., Papadaki, M., Bak, N.: Design of the Blockchain smart contract: a use case for real estate. J. Inf. Secur. **9**, 177 (2018)
13. Miles, C.: Blockchain security: what keeps your transaction data safe? IBM Blockchain. https://www.ibm.com/blogs/blockchain/2017/12/blockchain-security-what-keeps-your-transaction-data-safe/ (2018)
14. Salman, T., Zolanvari, M., Erbad, A., Jain, R., Samaka, M.: Security Services Using Blockchains: A State of the Art Survey. IEEE Commun. Surv. Tutor. **21**(1), 858–880 (2018)
15. Szabo, N.: Smart Contracts: Building Blocks for Digital Markets. http://www.fon.hum.uva.nl/rob/Courses/InformationInSpeech/CDROM/Literature/LOTwinterschool2006/szabo.best.vwh.net/smart_contracts_2.html (1996)
16. Wang, S., Ouyang, L., Yuan, Y., Xuan, X., Wang, F.: Blockchain-Enabled Smart Contract: Architecture, Applications, and Future Trends. In: IEEE Transactions on Systems, Man, and Cybernetics, pp. 1–12 (2019)
17. Stark, J.: Making Sense of Blockchain Smart Contracts. https://www.coindesk.com/making-sense-smart-contracts (2016)
18. Ethereum. https://www.ethereum.org/
19. Solidity. https://solidity.readthedocs.io/en/v0.5.3/
20. García, N., Peiró, E.J.: Martinez, Blockchain and Land Registration Systems. European Property Law Journal **6**, 296–320 (2017)
21. Veuger, J.: Trust in a viable real estate economy with disruption and blockchain. Facilities **36**, 103–120 (2017)
22. Crichton, D.: With MIT Launched, Learning Machine raises seed to replace paper with blockchain credentials. https://techcrunch.com/2018/05/07/learning-machine-credentials/ (2018)
23. Pelletier, M.: What is Blockchain and How Can it be Used in Education. https://mdreducation.com/2018/08/20/blockchain-education/ (2018)
24. Li, R.: Better Security Over Blockcerts, MSc Thesis, School of Computer Science, University of Birmingham. https://dgalindo.es/mscprojects/rujia.pdf (2017)
25. Blockcerts. https://www.blockcerts.org/
26. CreativeCommon. https://creativecommons.org/
27. Wemark Image Licensing. https://www.wemark.com/ (2019)
28. Copyrightbank, Copy Right Registration and Verification. https://www.copyrightbank.com/ (2019)

IoMT: A Blockchain Perspective

Raluca Maria Aileni and George Suciu

Abstract This chapter presents several aspects concerning the evolution and importance of the Internet of Medical Devices (IoMT) applications and blockchain technology in healthcare for patients' data analysis and research about the adequate medication for diseases such as cancer, Alzheimer or epilepsy. Blockchain can help in creating a single database to collect data during clinical trials and allow patients' data security. In the healthcare system, blockchain is based on P2P (peer to peer) network computers (nodes), which can optimize the efficiency of the Internet of Things (IoT) and can ensure a safe connection of the medical equipment. According to statistics, spending on blockchain solutions worldwide from 2019 to 2022 is increasing from $2.9 billion to $12.4 billion. The novelty of the chapter consists of providing a comparison between cloud computing and blockchain, together with a connection between two of the essential radical innovation breakthroughs, i.e. blockchain. In this chapter, there are also presented the IoMT applications based on blockchain for healthcare and aspects concerning data security in blockchain.

1 Introduction

Healthcare represents a data-intensive clinical area where massive amounts of data are produced, accessed, as well as significantly challenging, due to the sensitive nature of data and restraining factors, such as security and privacy [1]. In the healthcare domain and clinical frameworks, safe, secure, and scalable (SSS), data-sharing is paramount for diagnosis, as well as in linked clinical decision making. The data-sharing practice is considered necessary to allow clinical practitioners to transfer the clinical data of their patients into a privacy-sensitive and appropriate manner, ensuring that both participants have exhaustive and up-to-date information about patient health conditions.

Therefore, telemedicine and e-health represent two areas, which are intensively used for clinical data to be transferred remotely towards a technician at a remote site, for example, asking an experts' opinion. Through these clinical services, the

R. M. Aileni (✉) · G. Suciu
Politehnica University of Bucharest, Bucharest, Romania

M. A. Khan et al. (eds.), *Decentralised Internet of Things*, Studies in Big Data 71,
https://doi.org/10.1007/978-3-030-38677-1_9

patient's data is transported either by a "store-and-forward" technology or using real-time clinical monitoring, for instance, telemonitoring and telemetry [2]. Using these clinical services, the patients are remotely diagnosed and operated by clinical specialists using transacted clinical data. Throughout these clinical systems, the security, sensitivity, and privacy of data are some of the most significant features to materialize, due to the sensitive nature of patient data. Hence, the capacity to exchange data safely, securely and scalable is very relevant to sustain reliable clinical communications, concerning remote patient cases [3, 4]. Throughout the past several years, researchers have worked to implement applications for the Internet of Things (IoT), artificial intelligence (AI), machine learning (ML), and computer vision to expedite doctors and clinical practitioners in the diagnosis and surgery of numerous persistent diseases.

Blockchain has a wide range of applications, but the new one that has started to receive attention on blockchain applications in the healthcare domain [5]. The blockchain technology is designed to establish trust, accountability, traceability, and integrity of data sharing, which are very important for a healthcare system. In the IoT healthcare systems, one of the goals is to secure distributed data across an organization or even national boundaries, which are insured by blockchains. Healthcare is a system composed of many entities, like patients , medical personnel, healthcare providers, treatment or equipment providers, and educational or research institutions. General aspects that are being pursued to use blockchain in healthcare systems are the storage of purchased data with various medical devices and equipment, data security, information confidentiality, integrity protection, availability, accuracy, and data compliance of individuals [6]. From a confidentiality perspective, blockchains can be accessed by parties interested in a particular network, and access can be controlled by proper security policies (role-based or attribute-based). By ensuring and protecting the confidentiality of sensitive information, only authorized parties should be granted access to sensitive data stored in the blockchain. Using blockchain and protecting data integrity from healthcare involves protection against unauthorized data deletion or modification. The access time for data in healthcare is critical to this area. Thus, a system based on blockchain and medical devices should also ensure a fast response time. Technologies and devices used in healthcare must not be exposed to attacks, and classic systems require improvements to avoid these attacks, so many solutions are being developed in this area. Unfortunately, there have been countless cases in which patient data has not been protected. For example, in 2013, Kaiser Permanente (USA) informed patients that health information was compromised by the theft of an unencrypted USB flash drive containing medical records of patients [7].

The chapter structure is separated into 6 sections. The current section represents the introduction, having the primary purpose of providing sufficient information for a general idea of the blockchain and healthcare domains. In Sect. 2, it is presented a description of how IoMT can benefit the healthcare sector. Several opportunities of the blockchain technology within healthcare are presented in Sect. 3. A parallel discussion on BlockCloud cloud computing in comparison to a blockchain is undertaken in Sect. 4, while Sect. 5 brings a connection between AI algorithms and the blockchain architecture. Section 6 is more focused on security and data privacy through blockchain. The last section of this chapter concludes and envisions future work.

2 IoMT Applications for Healthcare

The Internet of Medical Devices (IoMT) plays an essential role in the healthcare industry in improving the accuracy, reliability, and performance of electronic devices. Researchers contribute to the digitization of the healthcare system by combining available medical resources and services. This chapter is a brief report on the implementation of the IoT and its application in the field of healthcare.

The Internet of Things in the field of healthcare focuses on the following:

- Critical treatment of severe threats to life;
- Medication and patient' routine examination;
- Standard critical processing and connection to machines, people, and data that can be displayed on a machine or in the cloud.

The current use of IoT in healthcare can be divided into two categories [8]. The first is to identify, monitor, and maintain assets, and the second is to improve the efficiency of service levels. Remote monitoring provides easy access to current health status, thanks to key wireless solutions connected via operating system interface to monitor patients with registered and protected health status data. Several sophisticated algorithms and sensors are used to analyze the data, which is then transmitted via wireless to follow medical recommendations. Clinical care uses non-invasive IO (input-output) monitoring systems for hospitalized patients. This clinical management system uses sensors to collect physiological information that is stored and analyzed in the cloud. It provides an automated, continuous information flow system that improves the quality of medical care at a lower cost.

2.1 Devices in Health Monitoring

While most patients need to commute to the hospital, through the Internet of Things, they can receive remote treatment from their comfort zone. Healthcare technologies support many features, such as real-time integration of health data with monitoring useful medical and non-medical data flow and electronic medical records. Referring to the aged people, their regimen includes Internet of Things ultrasound technology, used in hospitals as a customized home care solution to locate and monitor the activities of residents. Alarms can be handled in a real cost system for a comprehensive communication interface. For example, a system of used and sealed sensors can be programmed to send a position report to the ultrasonic receiver. The receiver, in turn, receives the signal via typical wireless WLAN (Wide Local Area Networks)

Table 1 Health monitoring examples

Health monitoring	Parameters
Patient's vital signs monitoring	Hearth rate, blood pressure, body temperature, glycemia, conductivity, respiratory rate, SpO_2
Patient's surroundings	Temperature, relative humidity, air pressure, air quality (PM 0.1–0.3), luminosity, noise level
Early warning system for doctors/medical staff	Signals sent via email/messaging apps to the doctors/medical professionals
Context-aware applications	Saved settings for an identified individual

connections to the residential gateway. Usually, the data analysis is carried out by the gateway, sending essential data, and the integrated wireless connection is used to send service notifications in the event of critical events. The Internet of Things will become more and more present in the medical field. Even if the implementation starts significantly, this area is still in the integration phase. Some sectors to be monitored, along with examples from each, are presented in Table 1.

Many portable devices have been installed to measure the critical elements of health monitoring data. Recently, efforts have been made in the field of portable devices for the measurement of multitasking of vital data. Many devices, projects, developments, and solutions for portable remote ECG monitoring, which plays an essential role in health monitoring have been proposed in the literature and industry. These solutions are often difficult to implement and are not sufficiently effective in terms of energy consumption or efficiency. Some of them are notable but cannot be merged with other signals from different systems.

2.2 *Devices in Healthcare Improvement*

An example of an IoT solution was developed for diabetic retinopathy. Diabetic retinopathy is a severe disease that is the leading cause of adult blindness. The only key to prevent this disease is to detect it early in the screening process. Unfortunately, only 40% of patients are screened. To facilitate early diagnosis, Pensacola's Intelligent Retinal Imaging System (IRIS) has developed a screening solution available in rural and urban areas with an automatic camera [9]. The answer is to send retina images captured by the automatic camera to the cloud, that has an IRIS algorithm designed to detect diseases. The diagnostic report is returned within 24 h.

There are a variety of possibilities in this domain, such as:

- Implantable medical devices, such as pacemakers, defibrillators, and neurostimulators;
- Concussion detection in sports;
- Helmets, patches and mouth guards;

- Motion detection and body motion reconstruction;
- Man-down and personal emergency response systems;
- Rehabilitation and training;
- Improved straight-line motion and tilt detection for safety;
- Instrument guidance in surgery;
- Healthcare mobility aids, including wheelchairs and scooters.

3 Opportunities for the Blockchain Technology in Healthcare

Any healthcare blockchain would need to be public and include technological solutions for three key elements: scalability, the security of access and privacy of data. A distributed blockchain featuring health records, documents, or images would have implications for data storage and limitations for data throughput. Therefore, each user would have a copy of every health record, for example, that is located in the distributed healthcare blockchain network, this not being practical from a data storage point of view. The blockchain will act as an access control manager for health records and data.

Blockchain has proven to be a flexible solution in different market sectors because of network resilience improvements and the security of supply. The importance of blockchain can be highlighted through some applications in the healthcare industry. The advantages are visible, especially when different people need access to the same data. For example, in the case of chronic diseases, there are cases when a therapist or other medical specialist can face an incompatibility between Information Technology (IT) interfaces or other medical health records. If a problem occurs, it can lead to time-consuming authentication. Therefore, a decentralized database might produce numerous advantages in terms of accessing the same information. For example, a start-up called Gem launched Gem Health Network [10], an infrastructure based on Ethereum Blockchain technology that helps specialists to access the same data and will resolve other operational problems. Therefore, medical mistakes such as outdated information, will be diminished. Furthermore, the confidence level between the doctor and the patient will not be neglected, and the entire way of treatment will be characterized by transparency.

The demand for universal health apps that cope with large amounts of data has led to an interest in related medical researches, blockchain offering several ways to monitor parameters that might indicate some health problems. Swiss health start-up called HealthBank handles data transactions and the way personal data are shared through a platform where the information is securely stored and managed. Nowadays, people use HealthBank to save data from the platform and to use them for further medical research. HealthBank can be seen as a symbol of an end-user that will use the Blockchain technology to increase its ability to solve problems in this industry.

When it comes to a delicate production process, blockchain can be successfully used to monitor the production of medicines. The main problem, according to the World Health Organization is that, although most of them contain the active ingredient, the dose is either too high or too low. Counterfeit Medicines Project [11], launched by Hyperledger, focuses its research on drug counterfeiting by marking each product with a timestamp. Blocks encompass records of transactions. This kind of information structure allows provenance, i.e., a zone of origin for any transaction. Because every blockchain transaction is timestamped and permanent, fraudulent drug dealers can be effectively identified. This way, the exact details where a drug is being produced can be determined. By using Blockchain technology, the origin of the product and its components can be precisely known.

Nuco represents an example of a company that battles drug fraud through blockchain. Nuco created a system that will monitor medical prescriptions and will send an alarm through a monitor program that will improve access to a database and also the response time as it will scan data marked on the prescription. Nuco will be the intermediary between the doctor and the pharmacist and will provide complete feedback from this chain. Moreover, other companies, like HealthChainRx and Scalamed, came up with similar solutions.

Electronic Health Records (EHRs) represent another aspect to be taken into consideration for traceability when referring to blockchain advantages. MedRec management system [12] is designed to handle those records. Blockchain technology can solve the interoperability problems caused by the data that is exchanged between different business parties. Data sharing in healthcare brings new challenges in terms of security. Therefore a new approach, for instance, one based on the blockchain can play an essential role in improving the data sharing process. Interoperability can enhance communication systems and reduce the time spent working on administrative tasks. Hospitals and private clinics are the main business entities that will benefit from a well-structured system.

Gathering all data together can be a difficult task due to the high amount of information. By using blockchain, all the related data is securely collected. In this way, doctors will have a broader perspective of someone's medical history and will have uncomplicated access to the high amount of information. Another facility provided by this technology is related to the patient's identity. Blockchain uses public-key infrastructure (PKI) as an identification method.

The paper [13] detailed a "Healthcare Data Gateway" (HDG) that allows patients to supervise their data. However, the importance of a better understanding of the privacy and security implications that come with this technology is also highlighted.

The project entitled Medicalchain described in [14] aims to enhance the process of freeing up beds for the next patient and to make sure that all the documents that belong to each person are free of errors. Therefore, a digitized solution based on blockchain technology intents to reduce the time spent discharging the patient.

The dental industry also faces some problems nowadays due to the high number of independent doctors. Dentacoin is a platform which uses blockchain technology to connect the practitioners, the patients, and also the suppliers. Cryptocurrency is also used in this system as the patient can benefit from additional dental services if, for example, they write a review.

There are some cases where the importance of tracking becomes crucial. Detailed facts about pharmaceutical provenance can be integrated into a blockchain solution as it provides an efficient way to enter complete information.

Nowadays, telemedicine provides a facility that takes care of the patients who are situated in distant regions, far from local health services where doctors are in short supply. Therefore, patients who benefit from these services can avoid waiting in a doctor's office and will receive instant therapy. Telemedicine services are generally fitted with more sophisticated technology and far more extensive than traditional physical health services. By using blockchain technology, there is no need for a third-party power. However, blockchain technology cannot solve the complicated challenge of information sharing as it must be integrated into current disparate health systems and norms of clinical information.

Another application is related to the cancer diagnosis problem. Getting new insights from distinct experts can shorten the time for a patient with cancer from diagnosis to therapy. Today, most hospitals have included at least one tumor board where a team of medical experts will discuss each patient's condition and treatment options. A large business hospital may require experts from a broader spectrum of fields, while a smaller care facility may have restricted resources to expand their tumor boards. The existing health systems for cancer patients lack a patient-controlled data sharing function to request a second opinion quickly and also share information selectively. Blockchain technology provides the chance for trustless exchange and disintermediation that permits the aggregation of current relationships across different organizations and suppliers. In cancer care, data sharing is particularly critical [15].

4 BlockCloud: Cloud Computing vs. Blockchain

In this section, we analyze the convergence of cloud computing and blockchain technologies for IoMT.

4.1 Blockchain and Cloud Security

Cloud computing can allow medical staff and patients to store their data in the cloud and at the same time, give them access to applications or services within a pool of configurable computing resources. Data storage in the cloud could create severe problems concerning security and data privacy if data is not anonymized previously. Although the security of the cloud computing system and network influences the safety of the external data in the cloud, the cloud's main features as on-demand services, continuous network access, resource pooling, and elasticity are susceptible to security threats. Moreover, the leading cloud computing technologies for virtualization, cryptography, and web services have vulnerabilities due to insecure implementation. At the same time, in the cloud computing environment, the security controls such as key management, encounter several challenges. For example, implementing a key management system within the cloud infrastructure requires administration and storage of different types of keys. Because the virtual machines usually have heterogeneous hardware and software, the assignment of crucial standard management is a complicated process.

Currently, PKI-based signatures are used to protect data exchange within the cloud infrastructure. Specifically, to detect unauthorized data changes and to identify the responsible entity, a stronger attribution is needed. Information related to modifications performed on data exchanged between multiple entities can be extracted from data provenance. To ensure provenance, researchers have proposed security solutions, such as PKI signatures. Within the cloud infrastructure, the implementation of PKI signatures typically depends on a centralized authority, which is not practical.

As an alternative to PKI signatures, Blockchain and keyless signatures have been proposed. Blockchain technology facilitates the secure transfer of information using a sequence of cryptographically secure keys across a distributed system. A central authority is not needed since a system of distributed ledgers executes it. This system records all actions performed on the data, and it is shared among all the participating entities. In the public ledger, the transactions are verified by a consensus of the majority of participating entities. The blockchain contains records of transactions that cannot be changed. The issue of "PKI key compromise" is addressed by the keyless signature by separating the identifying signer and integrity protection processes from others that are responsible for maintaining the security of the keys.

The cryptographic tools chosen from options such as asymmetric or keyless cryptography are responsible for identifying signer and integrity protection processes. For example, one-way collision-free hash functions are included in keyless cryptography. Examples of keyless signature processes are hashing aggregation and publication. To create a keyless signature, a Keyless Signature Infrastructure (KSI) is required. KSI consists of a hierarchy of co-operative aggregation servers that generate global hash trees. The verification process within KSI is based on the security of hash functions and also on the availability of a public ledger (blockchain). The blockchain is publicly available, and rules for updating, distributed consensus, and mode of operation are strictly defined.

Guardtime has proposed changes to traditional blockchain technology regarding integration with KSI. To mitigate challenges associated with mainstream blockchain technologies, KSI blockchain technology was developed. These challenges include a lack of scalability, consensus, and lack of formal security proof [16].

4.2 Blockchain and Cloud Security

With the rapid increase in the amount of data, the capacity of traditional storage devices cannot meet today's demands. In this regard, the Cloud Storage Systems (CSS) have been improved to increase storage capacity. CSS is divided into three categories, depending on the type of data stored: Block-based Cloud Storage Systems (BCSSs), Object-based Cloud Storage Systems, and File-based Cloud Storage Systems. Among these Cloud Storage Systems, the BCSSs are the most popular at the moment, because they have the lowest cost and the easiest scalability. Within BCSSs structure, data is stored in a block structure, so the resources consist of small block spaces. When a user submits a data file to BCSSs, the systems split the file into segments, and then transform those segments into property blocks [17].

4.3 Blockchain Based Healthcare Applications

New healthcare technologies based on blockchains are conceptually organized into the following four levels: data sources, the blockchain technology itself, health applications, and stakeholders, as shown in Fig. 1 [18].

This workflow has also got four primary levels, including raw data on healthcare, a chain of custody technology, health applications, and stakeholders. The blockchain is a distributed technology that allows multiple stakeholders to benefit from healthcare applications management [19–21].

Firstly, all data from medical devices, laboratories, social media, and other sources are compiled and processed into raw data, which is then scaled up to detailed data. These data represent an essential component of all healthcare in the chain and also the

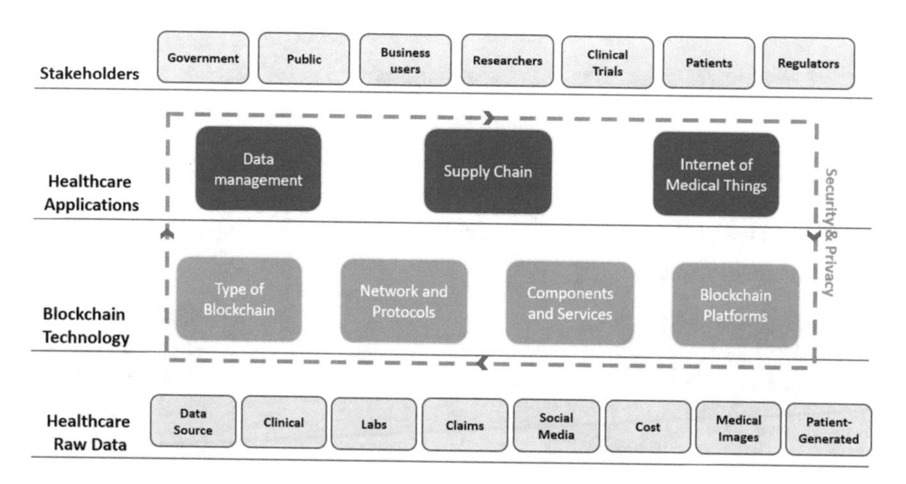

Fig. 1 Blockchain based workflow for health applications

main element that forms the first layer of the workflow. The blockchain technology is at the top of the raw data layer, which is considered the central platform for creating a secure healthcare architecture that is divided into four components.

Each platform in the blockchain has different properties, such as algorithms and consensus protocols. Several blockchain platforms have been created, such as Ethereum [22], Ripple or Hyperledger Fabric. The main components of the blockchain are smart contracts, signatures, portfolio, events, membership, and digital assets. Various protocols can be used to communicate with other programs and frameworks, or even between multiple networks.

Once the platform is created by deploying blockchain technology, the next step is to ensure that the applications are integrated into the global system. Healthcare applications based on the blockchain can be divided into three principal classes: data management, including the exchange of comprehensive scientific data for research and development (R&D), data storage (e.g., cloud applications) and electronic health records (EHR) [23]. The second class covers supply chain management (SCM) applications [24, 25], including clinical and pharmaceutical studies. Finally, the third class covers IoMT (Internet of Medical Things), which includes the interaction between the Internet of Medical Devices (IoMT) and data security [26, 27]. Figure 2 illustrates several healthcare applications in the blockchain.

Each transaction in a chain of health-related blocks is recorded in a distributed storage system. In a healthcare system, medical patient data is organized into an EHR, which is considered the fundamental component of a large distributed medical repository. These can be stored on-site or in the cloud, with security being a priority. Cloud storage is mainly the composition of many storage devices connected to a large volume of memory to accommodate a large amount of IT infrastructure. A healthcare system based on a chain of blocks is an example of such an IT infrastructure. Cloud storage technology offers the advantages of fast transfer, good sharing, storage capacity, low cost, easy access, and dynamic allocation. A patient-centered

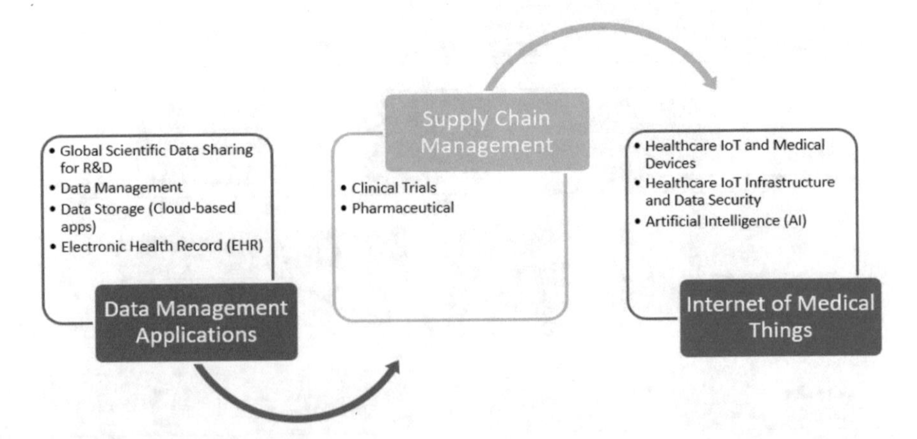

Fig. 2 Blockchain based healthcare applications [15]

health data management system can be proposed in a cloud environment that uses blockchain technology as a storage method to preserve confidentiality. The main idea of this work is to keep sensitive health data in the blockchain by defining a set of security and confidentiality requirements to ensure accountability, integrity, and security.

Thus, a new term has been introduced: BlockCloud, which is a mixture of blockchains implemented in a cloud environment. Cloud implementation has the idea of keeping distributed and secure data under the same roof without involving third parties [28].

Consideration must be given to how medical providers and organizations, public health organizations, health service providers, and governments should collaborate and create mechanisms for policy implementation.

5 Blockchain Architecture for Healthcare Applications

In this section, we evaluate blockchain architectures for healthcare applications.

5.1 Blockchain Architecture in Healthcare

Blockchain is an architecture platform which works as a distributed database that is used to maintain a growing list of records [5]. These records consist of blocks that lock with each other through specific cryptographic mechanisms. Typically, the blockchain is supported by a network of peer-to-peer users who collectively accept the previously established rules for the acceptance of new blocks. Each block record contains a time or signature stamp and a link to a previous block in the string. The design of the blockchain ensures that the data remains the same. As a result, the data of a particular block cannot be changed later without changing all subsequent blocks, and the consent of all members of the network. Its integrity and durability allow the blockchain to be used, distributed and compromised across different components or database network systems in an efficient, verifiable, and consistent manner.

Anyhow, blockchain has vulnerabilities that are specific for the system's architecture as blockchain specific attacks, scalability, and mining incentives. The purpose is to define an architecture that can solve at least one limitation of the blockchain infrastructure because healthcare data is crucial. The architecture presented in Fig. 3, created from a Wireless Sensor Network (WSN), allows the blockchain to be used optimally.

The system architecture, which facilitates the blockchain technology integration includes the following entities:

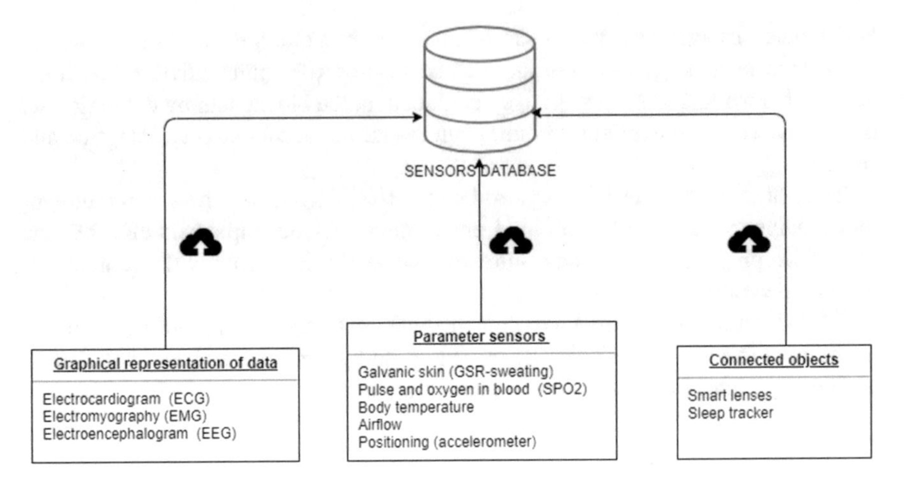

Fig. 3 Personal data collection

6 Blockchain Security and Data Privacy

Referring to the general architecture of IoMT, it is composed of three layers [29]: the perception layer, the network layer, and the application layer, as described in Fig. 4. The main task of the perception layer is to assemble health care information with the different type of devices. The network layer, which is made of wired and wireless systems and middleware, operates and transfers the input gathered by the perception layer sustained by technological platforms. The application layer integrates the medical data sources to implement personalized medical services and comforts the

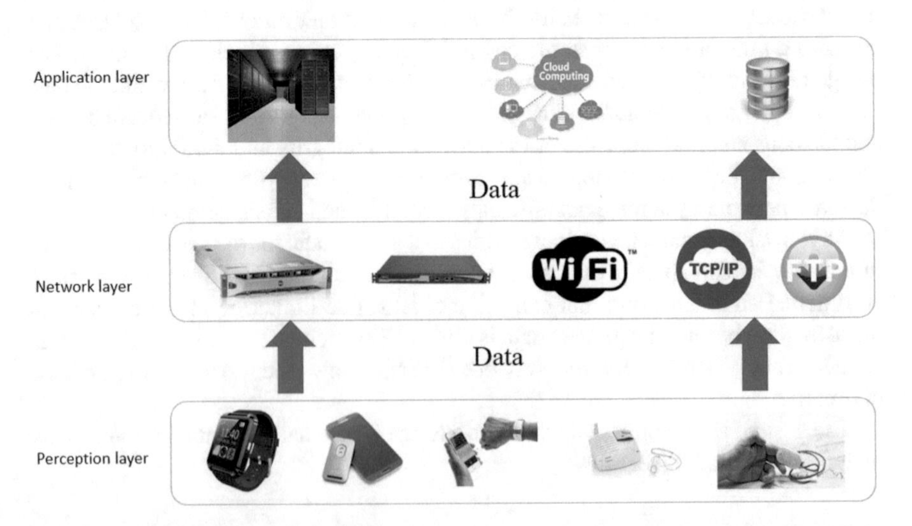

Fig. 4 IoMT architecture

definitive users' needs, according to the actual situation of the target community and the service demand.

All the data gathered from the devices presented in the architecture is stored in different databases. The traditional problem of synchronizing distributed databases is solved by combining peer-to-peer networks and using a distributed consensus algorithm. Public-key cryptography stands at the base of blockchain security. The tokens transmitted within a network belong to the address represented by a public-key.

Also, the blockchain data is considered incorruptible.

The blockchain's main work processes are as follows [30]:

1. The sender node records the new data and then broadcasts over the network;
2. The receiving node checks the message of the received data, and if the message is correct, the data will be stored in a block;
3. The network receiving nodes must execute the Proof of Work (PoW) algorithm on the block;
4. After the execution of the consensus algorithm, the block will be stored in the chain, all nodes in the network will accept the block, and the base of the chain will continuously develop on this block.

With Proof of Work, the probability of mining a block is directly proportional to the work done by the miner (for example, to check hashes using CPU/GPU cycles) [31]. This mechanism can bring people together to extract more blocks and make the "mining pools" become a place where computing power is most significant. When a computing power of 51% is reached, it can take control of this blockchain, but security problems can occur. Let us take an example in the field of health.

Thanks to the data-intensive domain of healthcare, a large amount of data can be created, stored, disseminated, and accessed daily. Healthcare data contain sensitive information that may be of interest to cybercriminals. The data may be sold to a third-party provider, and through the performance of data analysis to identify individuals who are not insured (due to genetic disorders or medical history) so that cybercriminals can benefit financially from the theft of the data.

Healthcare data may have integrity and confidentiality characteristics and must, therefore, be protected from external and internal attacks (attempts at the access that are not allowed from the network or ecosystem). Attacks, such as data modification or leakage, maybe unintentional or intentional. Organizations may be penalized or criminally liable if such an incident occurs, for example, under the Health Insurance Portability and Liability Act.

EMR/EHR/PHR (Electronic Medical Records/Electronic Health Records/Personal Health Records) ecosystem protection and privacy and integrity protection are active research areas. Approaches include the use of cryptographic primitives based, for example, on critical public infrastructures and public clouds to ensure data confidentiality and confidentiality. However, this limits the ability to search for data, as service providers must decrypt the (potentially significant) data before searching for decrypted data, which increases the time, cost of data discovery and diagnosis (downloading, decrypting and searching).

Access control models have additionally been utilized to manage and restrict access to the information by using predefined policies of access. These models are prone to be adequate against outside attacks. However, they are commonly inadequate against inside hackers as they are probably going to have already the rights to access to the information. Additionally, there have been ways to include some cryptographic primitives like attribute-based encryption with access control.

It is essential to keep data confidential and processing performance high in distributed ledge technologies [32]. However, in the case of time-critical applications, the blockchain is not appropriate because the transmission time is long for the block to be included in the chain. With Bitcoin, this process takes about 10 min [33].

The blockchain will have to use significant resources in the field of intelligent healthcare, for example, by relying on the Internet of Medical Things to ensure security, privacy, and valid insurance.

7 Conclusions

In this book, the chapter was presented various aspects regarding the importance of IoMT applications and the evolution of these applications in the context of their integration with blockchain technology. To address IoMT applications, two main concepts were identified: e-health and telemedicine. E-health concept is a relatively new domain supported by electronic communication and processes, while telemedicine refers to remote diagnosis and treatment of patients using telecommunication technology.

In the field of these applications, data security, information confidentiality, integrity protection, availability, and accuracy are essential aspects. In this regard, to ensure the security of the patients' data and of information transmission, blockchain technology was analyzed. Several advantages of this technology contain flexibility in different market sectors through network resilience improvements and the security of supply, time-saving authentication, up to date information, transparency, patient-doctor confidentiality, safe monitoring of drugs, patients, wearables, data, etc.

Moreover, an analysis of the convergence for blockchain technology and cloud computing (BlockCloud), as well as AI, has been presented.

Even though information integrity and distributed access of blockchain provide open doors for medicinal services information administration, these equivalent highlights likewise offer difficulties which require further study.

The reliable information integrity property of blockchain states that any information, once cast in blockchain, cannot be changed or erased. Still, when a record represents healthcare data, this information would go under the security of protection laws, a considerable lot of them would not enable individual information to be forever kept.

Another viable issue represents the way blockchain stores healthcare information. The technology was initially intended to record information exchange. One concerns itself about whether the present exchange can be followed in reverse to the initial "deal." Healthcare information, for example, imaging and treatment plans, a rigorous search, is required. How well blockchain storage can adapt to the two prerequisites is still to be discovered.

To manage these difficulties, it has been proposed the off-chain storage of information idea, where data remains outside of the blockchain in a traditional or a distributed database, yet the data hashes are included in the blockchain. Therefore, medical data remains off-chain and still can be verified, secured, and deleted appropriately. Simultaneously, unchanging hashes of the healthcare information are put away on-chain for verifying the legitimacy and precision of the off-chain medical records.

This idea though still faces potential difficulties. Due to the increasing levels of data protection laws around the globe and the endeavors by protection magistrates to view metadata of individual information as personal data, it is difficult to state that the hashes of individual information to be considered as personal information. At that point, the entire discussion of whether blockchain is fit to store individual data may start from the very beginning once again.

Still, there remain open difficulties that require further examination. For instance, cross-border sharing of healthcare information where various and frequently contradictory jurisdictions exist may hide the advantage of blockchain's information sharing. Undoubtedly, the anticipation for a person's privacy depends on every nation based on the administration guidelines. In this way, future research on guidelines, standardization, and cross-border healthcare information strategies, including data retention and use, is critical and going to be undertaken.

Another potential issue that is insufficiently researched represents the capacity of the blockchain to store and process large amounts of information access transactions in a convenient time. With the expanding volume of transactions, the delay of mining blocks will grow significantly. Thus, our future research will be focused on discovering innovative techniques for the reduction of blockchain mining delays.

Acknowledgements This work has been supported in part by UEFISCDI Romania, and MCI through projects ESTABLISH, PARFAIT and I-DELTA, and funded in part by European Union's Horizon 2020 research and innovation program under grant agreement No. 777996 (SealedGRID project), No. 787002 (SAFECARE project) and No. 826452 (Arrowhead Tools).

References

1. Griebel, L., Prokosch, H.-U., Köpcke, F., Toddenroth, D., Christoph, J., Leb, I., Engel, I., Sedlmayr, M.: A scoping review of cloud computing in healthcare. BMC Med. Inform. Decis. Mak. (2015)
2. Bhatti, A., Siyal, A.A., Mehdi, A., Shah, H., Kumar, H., Bohyo, M.A.: Development of a cost-effective telemonitoring system for remote area patients. In: 2018 International Conference on Engineering and Emerging Technologies (ICEET) (2018)

3. Castaneda, C., Nalley, K., Mannion, C., Bhattacharyya, P., Blake, P., Pecora, A., Goy, A., Suh, K.S.: Clinical decision support systems for improving diagnostic accuracy and achieving precision medicine. J. Clin. Bioinform. (2015)
4. Zhang, P., White, J., Schmidt, D.C., Lenz, G., Rosenbloom, S.T.: FHIRChain: Applying blockchain to securely and scalably share clinical data. Comput. Struct. Biotechnol. J. **16**, 267–278 (2018)
5. Onik, M.M.H., Aich, S., Yang, J., Kim, C.-S., Kim, H.-C.: Blockchain in healthcare: Challenges and solutions. In: Big Data Analytics for Intelligent Healthcare Management, pp. 197–226 (2019)
6. Ribitzky, R., Clair, J.S., Houlding, D.I., Mcfarlane, C.T., Ahier, B., Gould, M., Flannery, H.L., Pupo, E., Clauson, K.A.: Pragmatic, interdisciplinary perspectives on blockchain and distributed ledger technology: Paving the future for healthcare. Blockchain Healthc Today (2018)
7. Kaiser reports second fall data breach. In: Healthcare IT News. https://www.healthcareitnews.com/news/kaiser-reports-second-fall-data-breach. Accessed 24 May 2019
8. Dey, N., Ashour, A.S., Bhatt, C.: Internet of things driven connected healthcare. In: Studies in Big Data Internet of Things and Big Data Technologies for Next Generation Healthcare, pp. 3–12 (2017)
9. From the person, to the cloud, and back: How remote patient monitoring and intelligent feedback loops are driving the personalization of care. In: Microsoft: Digital Transformation in Health. https://info.microsoft.com/rs/157-GQE-382/images/Remote%20Patient%20Monitoring%20WW%20Health%20FY18%20Q2%20Vertical%20GEP%20Acquiring%20Content%20v2.0.pdf. Accessed 27 May 2019
10. Bitcoin News. In: Bitcoin Magazine Bitcoin and Blockchain News, Prices, Charts & Analysis. https://bitcoinmagazine.com/articles/the-blockchain-for-heathcare-gem-launches-gem-health-network-with-philips-blockchain-lab-1461674938. Accessed 27 May 2019
11. Jamil, F., Hang, L., Kim, K., Kim, D.: A novel medical blockchain model for drug supply chain integrity management in a smart hospital. Electron. **8**, 505 (2019)
12. MedRec. In: MedRec. https://medrec.media.mit.edu/. Accessed 25 May 2019
13. Yue, X., Wang, H., Jin, D., Li, M., Jiang, W.: Healthcare data gateways: Found healthcare intelligence on blockchain with novel privacy risk control. J. Med. Syst. (2016)
14. Albeyatti, A.: White paper: Medicalchain. MedicalChain self-publication (2018)
15. Zhang, P., Schmidt, D.C., White, J., Lenz, G.: Blockchain technology use cases in healthcare. In: Advances in Computers Blockchain Technology: Platforms, Tools and Use Cases, pp. 1–41 (2018)
16. Blockchain for distributed systems security. In: Google Books. https://books.google.ro/books?hl=en&lr=&id=dhaMDwAAQBAJ&oi=fnd&pg=PA161&dq=blockcloud&ots=QT3fd8Nrlu&sig=VgGbar98NYZE6ZTeNoGV1GbtQ_s&redir_esc=y#v=onepage&q=blockcloud&f=false. Accessed 28 May 2019
17. Wang, H., Jin, Z., Zhang, L., Jing, Y., Cao, Y.: Reasoning about cloud storage systems. In: 2018 IEEE Third International Conference on Data Science in Cyberspace (DSC) (2018)
18. Khezr, S., Moniruzzaman, M., Yassine, A., Benlamri, R.: Blockchain technology in healthcare: A comprehensive review and directions for future research. Appl. Sci. **9**, 1736 (2019)
19. Pilkington, M.: Can blockchain improve healthcare management? Consumer medical electronics and the IoMT (2017)
20. Meng, W., Li, W., Zhu, L.: Enhancing medical smartphone networks via blockchain-based trust management against insider attacks. IEEE Trans. Eng. Manag. (2019)
21. Jahankhani, H., Kendzierskyj, S.: Digital Transformation of Healthcare. In Blockchain and Clinical Trial, pp. 31–52. Cham, Springer (2019)
22. Malamas, V., Dasaklis, T., Kotzanikolaou, P., Burmester, M., Katsikas, S.: A forensics-by-design management framework for medical devices based on blockchain. In: 2019 IEEE World Congress on Services (SERVICES), Vol. 2642, pp. 35–40, IEEE (2019)
23. Nguyen, D.C., Pathirana, P.N., Ding, M., Seneviratne, A.: Blockchain for secure EHRs sharing of mobile cloud based e-health systems. IEEE Access (2019)

24. Khezr, S., Moniruzzaman, M., Yassine, A., Benlamri, R.: Blockchain technology in healthcare: A comprehensive review and directions for future research. Appl. Sci. **9**(9), 1736 (2019)
25. Hussein, A.F, ALZubaidi, A.K, Habash, Q.A, Jaber, M.M.: An adaptive biomedical data managing scheme based on the blockchain technique. Appl. Sci. **9**(12): 2494 (2019)
26. Kendzierskyj, S., Jahankhani, H.: Blockchain as an efficient and alternative mechanism for strengthening and securing the privacy of healthcare patient and clinical research data. In: 2019 IEEE 12th International Conference on Global Security, Safety and Sustainability (ICGS3), pp. 212–212, IEEE (2019)
27. Nimra Dilawar, M.R., Ahmad, F., Akram, S.: Blockchain: Securing Internet of Medical Things (IoMT)
28. Shi, P., Wang, H., Yang, S., Chen, C and Yang, W.: Blockchain-based trusted data sharing among trusted stakeholders in IoT. Softw.: Pract. Exp. (2019)
29. Sun, W., Cai, Z., Li, Y., Liu, F., Fang, S., Wang, G.: Security and privacy in the medical internet of things: A review. Secur. Commun. Netw. (2018)
30. Esposito, C., Santis, A.D., Tortora, G., Chang, H., Choo, K.-K.R.: Blockchain: A panacea for healthcare cloud-based data security and privacy? IEEE Cloud Comput. **5**, 31–37 (2018). https://doi.org/10.1109/mcc.2018.011791712
31. Lin, I.C., Liao, T.C.: A survey of blockchain security issues and challenges. Int. J. Netw. Secur. **19**(5), 653–659 (2017)
32. Suciu, G., Nădrag, C., Istrate, C., Vulpe, A., Ditu, M.C., Subea, O.: Comparative analysis of distributed ledger technologies. In: 2018 Global Wireless Summit (GWS), pp. 370–373. IEEE (2018)
33. Puthal, D., Malik, N., Mohanty, S.P., Kougianos, E., Yang, C.: The blockchain as a decentralized security framework [future directions]. IEEE Consum. Electron. Mag. **7**, 18–21 (2018)

Legal Ramifications of Blockchain Technology

Akinyemi Omololu Akinrotimi

Abstract The Blockchain technology deals with, and assures the security of assets and information over its network. However, if a breach occurs in some of its security features, and financial crimes such as fraud or tax evasion occurs, the law will have to be evoked to take its full course. Also, the Blockchain technology and the processes that accompany its implementation, exemplify a major deviation from the existing state of affairs in most financial activities, especially in the exchange of a legal tender for goods and services. Blockchain applications present contemporary methods of initiating and embarking on financial transactions in ways that do not conform to current legal structures. As such, having the judiciary decide on punitive measures for blockchain related financial crimes will require a meticulous assessment and revision of the rules guiding legal practice. It may be necessary for most countries to establish a nexus between certain scientific modules and ratification processes, so as to ensure that the modus operandi of blockchain applications agree with the law. This chapter seeks to provide a broad view of the relationship between blockchain and the Law, its effects on legal activities and how the Law can be used as a tool of protection in blockchain related transactions.

Keywords Blockchain · Cryptocurrency · Internet of things · Issues · Law · Smart contracts

1 Introduction

Law is a result of deliberate and logical effort of humans, to set standards for living, based on issues evolving from social interactions and continual arbitration of dispute. Law can also be described as a set of instructions, aiming at the deterrence

The original version of this chapter was revised: The order of the author Akinyemi Omololu Akinrotimi's name has been corrected. The correction to this chapter can be found at
https://doi.org/10.1007/978-3-030-38677-1_12

A. O. Akinrotimi (✉)
Department of Computer Science, University of Ilorin, Ilorin, Kwara, Nigeria
e-mail: akinrotimiakinyemi@ieee.org

M. A. Khan et al. (eds.), *Decentralised Internet of Things*, Studies in Big Data 71,
https://doi.org/10.1007/978-3-030-38677-1_10

or methodical settlement of conflicts. For a law to be publically accepted and effectual, it must be able to comprehend the activities of a wide range of individuals, businesses and organizations. This includes the approval or prohibition of rights to entities, property preceded by principles of enforcement, and providing procedure for punishment, restitution or remedy. The determination of approval or prohibition of rights and principles of enforcement has been an enduring foundation upon which every legal system rests. There has always been a need to create laws, to oversee the establishment of mutual agreements, due to the unpredictable nature of humans, in order to provide a basis for determining punitive measures for the erring party. The involvement of the law in these mutual agreements gave them a formal characterization known as *contracts*. Conventional contracts can be referred to as deals associated with requirements which are enforced by the law and they are based on specific legal rules which are designed to reflect the intentions of the parties directly executing the contract, with a legal authority involved in enforcing the intentions [1].

2 Smart Contracts and Traditional Contracts

One striking difference between smart contracts and traditional contracts is that, an autonomous code helps to enforce "blockchain-based" smart contract obligations, while obligations related to traditional contracts are often enforced by the court of law, with the aid of law enforcement authorities. Also, the codes that govern smart contracts are often executed by the underlying network nodes on a Blockchain. Using smart contracts, parties can enter into contracts with each other as total strangers without any distance limitation. The Blockchain allows for trust between parties without the need for an arbitrator, and acts as an unalterable record of the contract [2]. Traditional contract law provides a flexible platform where parties are able to distribute or allocate risks. This has to do with rectification (changing contracts, rescission-unwinding agreements made, especially under duress or because of unacceptable deals, restitution-returning the parties to a former condition the parties were before the contract was made and enforcement-in which a court orders a contract to be executed). In order to ensure this sort of flexibility, and a general conjecture of the possibility of using traditional laws in supporting smart contracts, traditional laws may need to be augmented to align with the deterministic and automated features of smart contracts—which pose various legal and regulatory challenges, ranging from customer protection mechanisms to enforcement methods, and the risk of a rise in illegal activities [3].

3 Blockchain and Cryptocurrencies

The need for digitalization of property, in order to build a sustainable system for transactions such as smart contracts, gave rise to the use of Cryptocurrencies. Cryptocurrencies such as bitcoins and the ethereum are ideals of the mainstreaming of Blockchain technologies [4]. Cryptocurrencies are however unstable; in that, their use echoes and often agrees with general market conditions. The current emphasis on cryptocurrencies in the teaching syllabuses of higher institutions is an indication of a gradual acclimatization of the society to distributed legal transactions in thought and practice. A survey of the top 50 universities in the world found that 42% offer at least a class on blockchain or Cryptocurrencies [5]. Cryptocurrencies have however not been globally accepted as a means of transacting financial activities, however cryptocurrencies have been seen as a potential solution to financial transaction problems for countries with weak or undeveloped payment infrastructure or remittance systems. These potential solutions came in form of viable cryptocurrency exchange systems such as Ripple [6] and blockchain remittance networks such as Abra [7]. A Blockchain network is built on a decentralized topology concept and therefore, it does not have a focal point of failure. This makes it capable of withstanding the devastating effects of highly malicious attacks which often target the hub of the network. In addition, entries on the blockchain are built upon past transactions, making it possible to conduct a comprehensive audit trial. To further secure cryptocurrency transactions on the blockchain network, digital keys are made available to its users in order to execute exchanges or transactions, however one of the major challenges involved with the use of the blockchain technology is that, if a user's digital key is stolen, the perpetrator could use it to plunder the user's account or alter the information on it for malicious reasons. In the event of such an occurrence, there is need to evoke the law to provide punitive measures. Some of the counties where the use of cryptocurrencies has been legalized make use of Blockchain regulatory organizations in curbing the excesses related to the use of the technology. For instance, in the United States (U.S), the IRS (Internal Revenue Service) has been using a blockchain investigative tool known as "blockchain analytics" since 2015, and has hired a special criminal investigation team to hunt down tax evaders such as those who transfer assets overseas, just as they do with their regular foreign accounts except that they do not pay tax, thereby employing the permissiveness of cryptocurrencies in that respect. The IRS was able to conduct its investigations with much success as they ordered Coinbase [8] (A U.S. based a secure platform that makes it easy to buy, sell, and store cryptocurrencies) in 2018 to reveal to the IRS, the details of 14,355 of its users who had transactions of over 20,000 dollars between, 2013–2015. This shows that there are rules that must be obeyed and that can be broken in using cryptocurrencies, thus, necessitating the intervention of the law.

4 Fundamental Needs of the Legal System for Blockchain in Smart Contracts and the Use of Cryptocurrencies

The huge level of dependability of the blockchain technology on computerized processes, presents contemporary methods of initiating and embarking on financial transactions, in ways that do not conform to conventional legal structure and it is right to adjudicate that computer codes do not violate laws, rather, people do. This implies that, in ensuring that this new technology does not completely lack the "legal touch", lawmakers must be aware of the conduct that can be produced by codified blockchain mechanisms, and close up observed orifices which can serve as potential avenues that fraudsters can employ, in carrying out illegal activities. Involving the law in monitoring blockchain practices raises legal concerns which can been seen as cogent reasons why legal institutions must be involved in monitoring blockchain transactions [9]. Some of the issues that suggest the need for the legalization of the mode of operation for person-to-person, local and international blockchain based transactions include the following:

i. **Legal Responsibility**: The functioning of blockchains is difficult to control and stop. This raises various issues for consideration in terms of liability management. The use of cryptocurrencies is generally deemed as the next frontier of development in the financial sector of the world economy, and its current level of acceptance around the world, basically stems from the decentralized nature of the blockchain technology; which is solely relied upon by miners in verifying cryptocurrency-based transactions. Hence, the requirements to introduce new technical and legal standards cannot be eliminated in order to overcome the increasing security issues associated with the use of the technology. For instance if a dispute erupts as a result of a breach in trust and agreement regarding a transaction between two customers, who bears the legal responsibility of settling such a dispute?

ii. **Legal Authority**: Since blockchain severs are decentralized and can be distributed around the globe in the event of the occurrence or failure in the system, taking a proper action across national boundaries may be very difficult, resulting in jurisdictional disorientation. For instance, in the event that there occurs a fraudulent activity on a blockchain network involving the transfer of a huge amount of currency across international borders, and the culprit is on the run, what are the legal rules, that may be used as the basis for taking legal decisions in the pursuit and prosecution of such an individual, if apprehended? Also, the more decentralized a system is, the more difficult it will be to organize a law enforcement crew to crack down on criminals.

iii. **Legal Standing of Decentralized Autonomous Systems (DAOs) as Blockchain units**: The functioning of the DAOs follows laws that are written as comport codes, as such, the software used in operating the DAOs has to be licensed and there must be a scope for the license. No software is totally free of bugs, therefore, running DAOs on a blockchain software that is not monitored, can put users' transaction in jeopardy. The law has to be involved in making sure that warranties or indemnities are provided to the users of a blockchain in order to induce more confidence and trust [10]. The disclaimers or exclusion and liabilities that the DAOs rely on, must also be made known to the users. Also, warranties, indemnities, disclaimers, exclusions and liabilities developed must be made known to the users with the help of the judicial system. Then, it can be clear who or what is claimed against in the case of a legal dispute.

iv. **Legal Enforcement of Smart Contract**: Smart contracts are pre-written as, and governed by computer codes; therefore it may be difficult to analyze them, using the concepts of conventional contract law especially in legal issues related to offer and acceptance or certainty and consideration. For example, considering the case of the need for renewal of a lease for real estate, where an estate owner leased out part of his property to a client, using blockchain technology based smart contract, what happens if client refuses to abide by the stated terms of the contract or in other words fails to renew the lease and is still in possession of the property?

v. **Insurance Issues**: Once a smart contract executes either of the parties involved in the smart contract cannot modify it, as the smart contract code automatically executes and does not consider factors that were not pre-written in the smart contract code [11]. So, in the case where a customer has entered into an insurance deal with an insurance company through a smart contract, what happens in a situation where the customer claims to have lost more than the amount the smart contract automated system, delivered as a payout for certain losses? For instance in the case of a natural disaster such as an earthquake, where the smart contract code calculates the losses and the amount to be paid to a concerned customer, using some attributes of the earthquake, such as seismic data—it might happen that, the value of what was destroyed is greater than what the smart contract executed as an insurance payout to the customer. Now where the customer can prove that the level of the destruction was more and demands a higher compensation, it will be important for the law to wade in. However, how does the law handle such an issue given that the terms of the contract had previously been agreed upon by both parties, when they initiated the smart contract?

vi. **Unlicensed Law Practice**: Blockchain smart contracts constitute and illicit practice of the law in that they give a smart contract coder, autonomy, in creating laws and terms under which a smart contract can be executed without carrying legal authorities along [12]. This automatically takes away the legal administration of smart contracts away from the reach of licensed lawyers and places it at the hands of computer programmers who can tweak constitutional contract laws in favor of their own personal interests, one or both parties involved in the contract. As such, there arises a need for licensed lawyers to officially take responsibility in working with smart contract coders in ensuring that the rules governing any smart contract is in-line with the dictates of the law.

vii. **Human Coding Error**: Suppose a dispute arises a result of a mistake which involves a coder making a mistake that breaches the agreement of a smart contract while coding the smart contract and this leads to the loss of some huge amount of money or privileged on the side of one of the parties involved in the smart contract after it has been executed. In such cases, legal authorities have to be involved in settling the dispute [13]. The legal institution involved will face the challenge of vindicating the affected party, in employing the traditional use of law in unwinding a smart contract that has already been executed.

viii. **Changing Customer Requirements**: The law profession is bound to face an increasing need for legal advice from customers as time goes by, as clients involved in the use of the smart contracts will expect their attorneys to understand the technology behind the blockchain in order to be able to advice them properly when doubts arise or in case of disputes. Eventually, as lawyers key into this new way of practicing the law profession and get accustomed to providing legal advice to their clients, to assist them in the use of the blockchain technology, there is a possibility of competition as clients will tend to take their legal concerns from a less tech-savvy law consultant or firm to one with a better idea of how blockchain-based technologies such as smart contracts function.

ix. **Monitoring and Enforcing Legal Injunctions**: As the use of the blockchain technology-based smart contract increase, there is likely to arise, the need of creating legal bans and orders using smart contracts, however there is a limit to which the technology can handle issues of such nature. For instance, in a case where a restraining order is served to an individual through the use of a smart contract, how does the smart contract monitor the individual to ensure that it terminates at the exact time the person breaches the agreement if at anytime the person does? Lawyers and traditional courts will definitely have to be involved in such cases in order to find a way of terminating the smart contract and applying punitive measures on the offending party.

5 Admissibility Issues Relating to Blockchain Technology

Despite the wide acclamation of the blockchain technology in most parts of the world, the growth in the usage of the technology by individuals and organizations has been stunted due to the following reasons:

i. **Confidentiality Issues**: There exists a possibility of user privacy concerns, as some users may be reluctant in putting their information on the blockchain network despite the promise of information security by the technology. This may render smart contracts, for instance, unfitting for most individuals and organizations in replacing traditional legal contracts, especially in dealings that require a substantial level of confidentiality.

ii **Recurrent Fluctuation in the Value Of Cryptocurrencies**: A lot of people who would have been users and potential users of cryptocurrencies belief that using cryptocurrencies could lead to great financial losses due to a continual fluctuation in its value. As such, cryptocurrencies are mostly used by speculators looking for nippy cash flow, including individuals and organizations that have personal problems with government-backed currencies or criminals who want a black-market way of exchanging money.

iii **General Skepticism for the Open Ledger System**: A lot of people, especially organizations believe that the open ledger system—in which all the transactions carried out on the blockchain is distributed onto all the systems on the chain, constitute a great security risk to their confidential information [14]. It is a general believe that: in any distributed trust system, there are exist backdoor methods for creep in.

iv. **General Skepticism for the Distributed Mining System**: Since just about anybody can decide to become a miner, there is a general skepticism for the mining system of the blockchain. There is a generally believe that miners can indeed get access to confidential information from a transaction they are mining.

v. **Cost**: The higher the number of miners on a blockchain, the faster the transactions over the blockchain will be. As at present, it takes about 10 min for a transaction to go through a blockchain network. This is called "*The Block Insertion Delay*". As such, there is need for more miners on blockchains. Mining a blockchain can however be very expensive, both in data storage, bandwidth usage and in the energy required to maintain it. The cost of engaging in mining a blockchain has therefore discouraged a lot of individuals and organizations in going into the business [15].

vi. **Institutional Rules and Regulations**: Most institutions have rules and regulations that compel most of their members to behave according to some norms, thereby imposing sanctions on those who do not. Therefore if an institution kicks against the use of Blockchain technology, most of their members are not likely to adopt the use of the technology, even for personal use, as the institution could have indoctrinated them on the reasons why they should not use it. In reality, this is why some persons have steered clear of the use of the technology.

vii. **Cyber attacks**: Villains use Trojans, phishing and password hacking tools, to obtain authentication requirements such as private and public keys in attempting to access individual accounts of users on the blockchain for fraudulent purposes. In some cases, successful attacks have been carried out on electronic wallets and exchanges [16].

viii. **The Irreversibility of Cryptocurrency transactions**: It has been very difficult to use blockchain systems in combination with other conventional systems. For instance, Modern banking, is designed to be reversible, however once a transaction is completed on a blockchain, it cannot be reversed. This makes it difficult in merging conventional banking systems with the blockchain technology, creating an inflexible and insecure system.

ix. **Difficulty in creating "The People-to-Technology trust shift"**: The process of carrying out blockchain enabled transactions, does not totally exclude the need to place some certain level of trust in human-led systems. For instance, there will always be a need for the government to be involved outside the system, but particularly, users still have to trust the cryptography, the protocols, the software, the computers and the network of organizations that provide blockchain services. As such, it has been generally difficult to convince a lot of organizations to shift the running of their organizations from the hands of the usual middle-men organizations like banks, to the blockchain system. Although, blockchain providers are meant to be relied upon absolutely, however they are also often viewed as single points of failure, especially in cases where the security system of their central storage facility gets compromised, usually through a cyber attack.

x. **Lack of knowledge**: Most individuals and organizations are still oblivious of how the blockchain technology works and how it can be of benefit to them, not to mention the risks involved in using the technology [17].

xi. **Patent concerns**: Intellectuals such as musicians and fine-artists believe that their music albums or art works are not secure on the blockchain, since the information on the blockchain are usually dispersed over various systems, even though they are encrypted. They believe that nobody can be held responsible if their information is stolen. As such, they often prefer to hold a human publisher they know and can personally interact with, responsible for securing their work.

xii. **Individuality of blockchain networks**: Separate blockchains do not work together [18], as such, if two users are willing to exchange information using the blockchain technology, then they must be on the same blockchain network. This makes it difficult for some organizations to adopt the use of the technology.

6 Traceability Issues Relating to Blockchain Technology

Tracing fraudulent activities on the blockchain has become a major concern in most parts of the world where the use of the technology has been widely adopted [19].

Some of the challenges involved in this respect are as follows:

i. **Anonymity**: Cryptocurrencies such as bitcoins are unnamed as they do not have any enforcement of "Know Your Customer (KYC)" attached to them, even though each transaction can be publicly viewed, making it easier for fraudsters to conduct illegal trades [20].

ii. **Invisibility of flipping scammers**: "Get rich schemes" using bitcoins on blockchain technology networks have caused a lot of people to lose a lot of money while giving them a false hope of having more. Unfortunately most of them have remained untraceable because of the nature of the technology. These schemes are hard to recognize and the scammer often eventually makes off with the victim's bitcoins.

iii. **Inability to crack down on Cryptocurrency pyramid schemes**: These cheating schemes are even harder to recognize than the bitcoin-flipping example described above. They often make use of investment programs and multi-level marketing. The scheme is often designed to multiply a low investment by signing up additional members, using referral links. Before long, hundreds of victims join the scheme but at a later point in time, the main scammer gives up the scheme and the pyramid collapses with many members of the scheme losing a lot of investment [21].

iv. **Inability to track down Bitcoin phishing impersonators**: There are a lot of impersonators on social media, who use the bitcoin brand itself as a ploy gain a victim's trust and reliability. They often use promise of bitcoin to lure their victims into following a URL that successively attempts to download a malware-laden application. These malware are then used to steal vital information from the victim's computer system.

7 Cyber Forensic Methods for Blockchain Technology

With a high level of vulnerability of blockchain networks to cyber attacks and a creation of a very important roles for legal institutions to occupy in libeling offences on individuals involved in carrying out such attacks, there is a need for viable approaches of uncovering the details of fraudulent blockchain related cyber activities, with the use of cyber forensics methods, in order to provide legal justifications for the law to take its course. Cyber forensics combines elements of law and Computer Science to collect and analyze data from computer systems, networks, wireless communications, and store devices in a way that is admissible as evidence in a court of law [19]. In order to ensure the reliability of the blockchain network and its ability to survive in the corporate world, the application of comprehensive computer forensics is essential.

i. **Notable Research work by Philip and Diana Koshy**: Among the first researchers to find a crack in the wall, were the husband-and-wife team of Philip and Diana Koshy [22]. In 2014, as graduate students in McDaniel's lab at Penn State, they built their own version of the software that buyers and sellers use to take part in the Bitcoin network. It was specially designed to be efficient in downloading a copy of every single packet of data transmitted by every computer on the internet, carrying out transactions using, Bitcoins. If the data flowing through the network were perfectly coordinated, with everyone's computer sending and receiving data as frequently as the rest, then it might be impossible for fraudsters to tag various individual IP addresses to their bitcoin transactions. But there is no top-down coordination of the Bitcoin network, and its flow is far from perfect. The Koshy's noticed that sometimes a computer sent out information about only one transaction, meaning that the person at that IP address was the owner of that Bitcoin address. And sometimes a surge of transactions came from a single IP address—probably when the user was upgrading his or her Bitcoin client software. Those transactions held the key to a whole backlog of their Bitcoin addresses. Once the Koshy's isolated some of the addresses, others followed.

ii. **Bitfury's Crystal**: According to Kyrylo Chykhradze, head of Bitfury's Crystal, this is a software solution designed to track activity on the Bitcoin and Bitcoin Cash Blockchains [23]. Crystal collects information about all the transactions recorded on a blockchain to determine which addresses belong to the same entity in order to be able to identify criminal activities, assigning a risk value to them, based on the type of activity.

iii. **Cogent research outcomes**: Many of the digital currency fraud incidences understudied in at different research centers have be linked back to the perpetrators. For instance, research activities have successfully traced 95% stolen bitcoins to fraudulent on-line cryptocurrency merchants.

8 Information Technology Laws for Blockchain Cryptocurrency Transactions

In Blockchain technology, Information Technology (I.T) Laws can help to provide legal approval for electronic transactions and also create a legal structure to help reduce cyber crimes. Table shows various countries of the world and the laws that apply to the use of cryptocurrencies in the various countries.

S/N	Country	Local law(s) related to blockchain cryptocurrency
1.	Australia	Australian Taxation Office (ATO) has published a guidance document on the tax treatment of virtual currencies According to the guidance, transacting with cryptocurrencies which is "akin to a barter arrangement, with similar tax consequences". Digital currency exchanges will be required to enroll in a register maintained by AUSTRAC (Australian Transaction Reports and Analysis Centre) and implement an AML/CTF program "to mitigate the risks of money laundering as well as identify and verify the identity of their customers" [24]
2.	Belarus	Individuals are permitted to engage in mining; acquire tokens; and exchange, sell, donate, bequeath, and otherwise dispose of cryptocurrency. Income generated by mining and operations in cryptocurrencies is exempt from taxation until 2023 [25]
3.	Canada	Digital currencies can be used to buy goods and services on the Internet and in stores that accept digital currencies. Individuals may also buy and sell digital currency on open exchanges, called digital currency or cryptocurrency exchanges [26]

(continued)

(continued)

S/N	Country	Local law(s) related to blockchain cryptocurrency
4.	Germany	Undertakings and persons that arrange the acquisition of tokens, sell or purchase tokens on a commercial basis, or carry out principal broking services in tokens via online trading platforms, among others, are generally required to obtain an authorization. It stated that firms involved in ICOs need to assess on a case-by-case basis whether the ICOs qualify as financial instruments (transferable securities, units in collective investment undertakings, or investments) or as securities and therefore trigger the need to comply with the relevant financial legislation [27]
5.	Israel	Although virtual currencies are not recognized as actual currency by the Bank of Israel, the Israel Tax Authority has proposed that the use of virtual currencies should be considered as a "means of virtual payment" and subject to taxation [28]
6.	Japan	Only business operators registered with a competent local Finance Bureau are allowed to operate cryptocurrency exchange businesses. The operator must be a stock company or a "foreign cryptocurrency exchange business" that is, a company, has a representative who is resident in Japan, and an office in Japan. A "foreign cryptocurrency exchange business" means a cryptocurrency exchange service provider that is registered with a foreign government in the foreign country under a law that provides an equivalent registration system to the system under the Japanese Payment Services Act. Cryptocurrency exchange businesses are required to separately manage customer's money or cryptocurrency apart from their own. The state of such management must be reviewed by certified public accountants or accounting firms [29]
7.	South Korea	Cryptocurrency dealers must have contracts with banks concerning cryptocurrency trades and the banks must examine dealers' management and cyber security systems before signing such contracts. In order to make a deposit into their e-wallet at a cryptocurrency dealer, a cryptocurrency trader must have an account at a bank where the cryptocurrency dealer also has an account [30]
8.	Malta	Malta, also known as the Blockchain Island, believes the potential of blockchain is endless. As part of their initiative to embrace this technology, the country has recently introduced two blockchain-related acts, known as Malta Digital Innovation Authority Act (MDIA ACT) and Innovative Technology Arrangement and Services Act (ITAS Act). MDIA Act focuses on certifying blockchain and establishing standards for the industry, while the ITAS Act focuses on setting up digital ledgers and regulating the entry of new blockchains [31]
9.	Slovakia	Virtual currencies must be treated as "short-term financial assets other than money". Cryptocurrencies directly obtained from mining shall be kept off-balance sheet until they are sold or traded [32]

(continued)

(continued)

S/N	Country	Local law(s) related to blockchain cryptocurrency
10.	United States	Under the Bank Secrecy Act (the "BSA"), a Money Services Business (MSB) such as a blockchain mining organization, is subject to the federal anti-money laundering regulations of the Financial Crimes Enforcement Network (FinCEN). The U.S Federal bankruptcy law requires a debtor who is declaring bankruptcy, to identify all the cryptocurrencies on the debtor's bankruptcy schedules. Upon the filing of a bankruptcy petition, such cryptocurrencies would become the property of the debtor's bankruptcy estate. If a debtor fails to disclose ownership of any cryptocurrency and the existence of such cryptocurrency is later discovered, this can result in: i. the bankruptcy court denying the discharge of the debtor's indebtedness or ii. in less frequent cases, the subsequent criminal prosecution of the debtor A cryptocurrency exchange is deemed to be a money transmitter that is subject to the same state licensing and regulation requirements as other money transmitters. In certain states (such as Colorado, Kansas, Pennsylvania and Texas), businesses engaging in certain types of cryptocurrency sales are exempt from state licensing requirements. Montana is currently the sole state that has no licensing requirements for money transmitters. Most states require money transmitters to submit to their money transmitter licensing and reporting through an account on the Nationwide Multistate Licensing System (NMLS) website. If a token issued in an initial coin offering (ICO) has utility, the token can be still be deemed to be a security that is regulated under the Securities Act. Every individual or business that owns cryptocurrency will generally need to, among other things, (i) keep detailed records of cryptocurrency purchases and sales, (ii) pay taxes on any gains that may have been made upon the sale of cryptocurrency for cash, (iii) pay taxes on any gains that may have been made upon the purchase of a good or service with cryptocurrency, and (iv) pay taxes on the fair market value of any mined cryptocurrency, as of the date of receipt [33]

Note Some of the countries not listed, are countries which either do not have specific legislation or regulatory laws as regards the use of cryptocurrencies or out rightly do not support the use of cryptocurrencies

References

1. Garry, G.: Policy considerations for the blockchain technology public and private applications. SMU Sci. Tech. Law Rev. **19**(3), Article 619-3 (2016)
2. Alexander, S.: Contract law 2.0 'Smart' contracts as the beginning of the end of classic contract law. Inf. Commun. Technol. Law J. **26**(2), 116–134 (2018)
3. Boucher, P.: How blockchain technology could change our lives. European Parliamentary Research Service. Available at: http://www.europarl.europa.eu/RegData/etud/idan/2017/581948/eprs_ida(2017)581948_en.pdf. Accessed 27 Jul 2019

4. Gupta, V.: The promise of blockchain is a world without middlemen. Harvard Bus. Rev. https://hbr.org/2017/03/the-promise-ofblockchain-is-a-world-without-middlemen (2017). Accessed 20 May 2019
5. n.a.: Coinbase research: 42% of top 50 universities offer at least one crypto-related class, Cointelegraph—the future of money. Available at: https://cointelegraph.com/news/coinbase-research-42-of-top-50-universities-offer-at-least-one-crypto-related-class (2018). Accessed 22 Apr 2019
6. Michael, C.: Forbes ripple deal could make xrp cryptocurrency compliant with FATF Anti-Money Laundering Rules. https://www.forbes.com/sites/michael-delcastillo/2019/06/26/ripple-deal-could-make-xrp-cryptocurrency-compliant-with-fatf-anti-money-laundering-regulations (2019). Accessed 17 May 2019
7. Yogita, K.: Crypto wallet Abra adds in-app support for 'Thousands' of US Banks. CoinDesk. Available at: https://www.coindesk.com/crypto-wallet (2019). Assessed 9 May 2019
8. Russell, B.: Coinbase ordered to report 14,355 users to the IRS: anyone moving more than $20,000 on the platform is subject to the new order. The Verge (29 Nov 2017). Available at: https://www.theverge.com/platform/amp/2017/11/29/16717416/us-coinbase-irs(2017). Assessed 9 May 2019
9. Ahmed, K., Andrew, M., Elaine, S., Zikai, W., Charalampos, P.: Hawk: the blockchain model of cryptography and privacy-preserving smart contracts. In: 2016 IEEE Symposium on Security and Privacy (SP), pp. 839–858, 2016
10. Andrea, P., Wiebe, R.: Distributed ledger technologies in securities Post-trading. European Central Bank, Occasional Paper No. 172 (2017)
11. Alan, C., et al.: Smart after all: blockchain, smart contracts, parametric insurance, and smart energy grids. Geo. Law Tech. Rev. **1**, 273–274 (2017)
12. Smart Contracts Application Examples and Use Cases. Draglet. Available at: https://www.draglet.com/blockchain-services/smart-contracts/use-cases. Accessed 22 April 2019
13. England/Australia: Goss v Nugent (1833) 110 ER 713; Mercantile Bank of Sydney v Taylor (1891) 12 LR (NSW) 252. In the US: Restatement (Second) of Contracts (1981) art 213; Uniform Commercial Code (UCC) art 2–202. A version of the parole evidence rule in the contractual context can be found in art 1341 of the French Civil Code, though other provisions do affect the manner by which proof may be leveled against parties involved in trade
14. Delmolino, K., Arnett, M., Kosba, A., Miller, A., Shi, E.: Step by step towards creating a safe smart contract: lessons and insights from a cryptocurrency lab, p. 2 (18 November 2015). University of Maryland. Available at: https://eprint.iacr.org/2015/460.pdf. Accessed 28 July 2019
15. Angela, W.: The Bitcoin blockchain as financial market infrastructure: a consideration of operational risk (2015) Legislation and Public Policy **18**, 837.12 Delmolino, Arnett, Kosba, Miller and Shi, above n3, p. 1
16. A recent Forbes article predicted global cybercrime to cost $2.1 trillion by 2019: Steve Morgan. Cyber Crime Costs Projected to Reach $2 Trillion by 2019. *Forbes* (17 Jan 2016). http://www.forbes.com/sites/stevemorgan/2016/01/17/cyber-crime-costs-projected-to-reach-2-trillion-by-2019/#4c1bba3bb0cc
17. Mayank, P.: Blockchain technology explained: introduction, meaning, and applications. Hackernoon. https://hackernoon.com/blockchain-technology-explained-introduction-meaning-and-applications-edbd6759a2b2 (2018). Accessed 2 Sep 2019
18. Most de Sevres N.: The blockchain revolution, smart contracts and financial transactions. Available at: https://www.dlapiper.com/en/uk/insights/publications/2016/04/the-blockchain-revolution (2018). Accessed 10 Oct 2018
19. United States Computer Readiness Team: computer forensics-overview. Available at: https://www.us-cert.gov/sites/default/files/publications/forensics.pdf (2008). Assessed 3 May 2019
20. Jose, P., Omri, R.: KYC optimization using distributed leger technology paper. SSRN Electronic Journal (2017). https://doi.org/10.2139/ssrn.289788
21. Vitalik, B.: Ethereum: a next-generation smart contract and decentralized application platform. Whitepaper (2014)

22. Koshy, P., Diana, K.: An analysis of anonymity in bitcoin using P2P network traffic. In: International Conference on Financial Cryptography and Data Security, pp. 469–485, 2015
23. Bitfury Group: Meet Bitfury: Bitfury launches crystal—a blockchain investigative tool for law enforcement and financial institutions. Retrieved from http://medium.com/meetbitfury/bitfury-lauinches-crstal-a-blockchain-investigative-tool-for-law-enforcement-and-financial-institutitons (2018)
24. Digital Currency, Financial Consumer Agency of Canada. Available at: https://www.canada.ca/en/financial-consumer-agency/services/payment/digital-currency.html. Last modified 19 Jan 2018. Archived at: https://perma.cc/G3PY-H8NR. Accessed 27 Jul 2019
25. Decree of the President of the Republic of Belarus No. 8 of Dec 21, 2017. Available at: http://president.gov.by/ru/official_documents_ru/view/dekret-8-to-21-dekabrja-2017-g-17716/. Archived at: https://perma.cc/S9UM-CKRG(in Russian). Accessed 27 Jul 2019
26. Digital Currency, Financial Consumer Agency of Canada. https://www.canada.ca/en/financial-consumer-agency/services/payment/digital-currency.html. Last modified 19 Jan 2018. Archived at https://perma.cc/G3PY-H8NR. Accessed 27 Jul 2019
27. Jenny, G.: Germany: Federal Ministry of Finance Publishes guidance on VAT treatment of virtual currencies. Global Legal Monitor (13 Mar 2018). http://www.loc.gov/law/foreign-news/article/germany-federal-ministry-of-finance-publishes-guidance-on-vat-treatment-of-virtual-currencies. Accessed 28 July 2019
28. Bank of Israel Capital Market, Insurance and Saving Branch, Tax Authority, Securities Authority and the Prevention of Money Laundry and Financing of Terrorism Authority, Public Announcement Regarding Possible Risks Contained in Virtual Coins. Available at: https://taxes.gov.il/About/PublicAnnouncements/PublishingImages/190214/oda190214.pdf. Accessed 27 Jul 2019
29. Takashi, N., Ken, K.: Development of legal framework for virtual currencies in Japan (Financial Services and Transactions Group Newsletter, April 2017). Available at: www.amt-law.com/pdf/bulletins2_pdf/160425.pdf. Accessed 27 Jun 2017
30. Financial Services Commission (FSC) [Special measures for the elimination of virtual currency speculation, enforcement of financial sector measures (28 Dec 2017)]. Available at: http://www.korea.kr/common/download.do?tblKey=GMN&fileId=185832583. Archived at: https://perma.cc/Z9YH-7SDM. Accessed 27 Jul 2019
31. Ivan, M.: Malta digital innovation authority unveiled: government working on green paper on artificial intelligence and internet of things. Times of Malta (16 Feb 2018). Available at: https://www.timesofmalta.com/articles/view/20180216/local/malta-digital-innovation-authority-unveiled.670847. Accessed 28 Jul 2019
32. Methodological Guideline of the Ministry of Finance of the Slovak Republic No. MF/10386/2018-721 for the procedure of taxing virtual currency. Available at: http://src.bna.com/xod(in Slovak). Accessed 30 Jul 2019
33. U.S. Digital Asset and Cryptocurrency Law. Retrieved 20 Jun 2019, from Berson Law Group LLP. Available at: https://blockchainlawguide.com (n.d.). Accessed 29 Jul 2019

On the Opportunities, Applications, and Challenges of Internet of Things

Shuvra Sarker, Kakon Roy, Farzana Afroz and Al-Sakib Khan Pathan

Abstract Over the course of time, the role of the Internet has changed a lot because of the evolution of various communication languages and billions of physical devices connected via the Internet around the globe. It is expected that in the near future, storage and communication services will be highly pervasive and distributed—in fact, we are already witnessing the trend to some extent. People, machines, smart objects, surrounding space and platforms connected with wireless/wired sensors, and many types of heterogeneous devices will create a highly decentralized resource pool brought together by a dynamic network of networks. Considering such a setting with millions of connected devices, in this chapter, we present a study on the best potential applications, challenges, and future opportunities in the area of Internet of Things (IoT). We discuss the general aspects and issues of IoT and explore the implication of all these in a developing country's setting taking the case of Bangladesh.

Keywords Applications · Architecture · Challenges · Internet of Things · Opportunities · RFID

1 Introduction

The Internet of Things (IoT) [1, 2] is a platform where various devices are connected to an Internet-like infrastructure so that they can interact and exchange data with each other. Essentially, it is an arrangement of interfacing physical things, interrelated registering gadgets, mechanical and computerized machines, articles, creatures or individuals that are given one of a kind identifiers, without expecting human-to-human or human-to-PC framework. With the help of IoT, the devices can interact, collaborate, and share experiences just like the humans do among themselves.

IoT has evolved from the concept of "*Web of Things*" that was initiated in around the year 2000 [3]. Over the long course of years, the concept has grown and increasingly, it is taking a more visible shape. It is expected that by next few years, we would be very close to implementing real-life, smart environments with the aid of

S. Sarker · K. Roy · F. Afroz · A.-S. K. Pathan (✉)
Department of Computer Science and Engineering, Southeast University, Dhaka, Bangladesh
e-mail: spathan@ieee.org

M. A. Khan et al. (eds.), *Decentralised Internet of Things*, Studies in Big Data 71,
https://doi.org/10.1007/978-3-030-38677-1_11

IoT technologies—at least we would be testing and developing usable IoT products with greater penetration into our daily *tech-life*.

In the recent years, we have witnessed significant changes in the communication languages among the electronic devices—we have got various types of interoperable protocols operating in heterogeneous environments and platforms. Storage and communication services have already become highly pervasive and distributed. It is expected that in the near future, people, machines, smart objects, surrounding space and platforms connected with wireless/wired sensors, and other types of electronic communications devices will get a more dynamic and decentralized environment interconnected by a dynamic network of networks.

While IoT equips a multitude of domains and millions of devices with connectivity, some domains of applications are relatively more interesting than others. There are in fact, many interesting and emerging technologies that would support IoT's operations. One of the attractive technologies that could be crucial for security purposes is Blockchain [4].

As the time is advancing, remote access and versatile registering are becoming relatively less expensive than that of the early period of such technology. There is a mechanism available which is called the omnipresent registering. While the registering centers are brilliant, savvy, and can be run with small contribution from the clients, the advancements in the electronics have made the portable gadgets relatively smaller in size than their predecessors. PDAs (Personal Digital Assistants), iPads, tablets and notepads supplanted normal mobiles and PCs. Subsequently, all these devices offer the users easy access to the web. In this scenario, when devices and technologies like sensors, Global Positioning System (GPS), actuators, etc. are added; the pervasiveness of the connected devices becomes stronger and greater. When the *crowd* (i.e., people carrying devices) are integrated in the environment, we get the setting like CrAN (Crowd Associated Network) [5]. In such a situation, gadgets are associated with the web (i.e., cyberspace) alongside the additional capabilities of sensing, processing, and performing canny undertakings. Overall, all these come within the greater definition of IoT which would have five basic components [6]: connectivity ("ubiquity of things"), object ID ("labeling things"), sensors and remote sensor systems ("feeling things"), inserted/embedded systems ("considering things"), and nanotechnology ("contracting things").

Given today's fast acceptance of new technologies and greater penetration into the populace, the amount of network traffic is growing at a remarkable speed. Cisco VNI (Visual Networking Index) forecasts 396 EB (Exabytes; 1 EB = 1 billion gigabytes) per month of IP (Internet Protocol) traffic by 2022 [7]. According to the conducted study, Global IP traffic would increase threefold over the period 2017–2022 and we are already witnessing it (at the time of writing this chapter). Overall, IP traffic would grow at a Compound Annual Growth Rate (CAGR) of 26% from 2017 to 2022. Monthly IP traffic could reach 50 GB per capita by 2022, up from 16 GB per capita in 2017. In this environment, security technology like Blockchain could be used to track billions of connected devices, for transaction processing and coordination among devices and to keep record of all transactions and data transfers. IoT industry could be really benefitted by the record keeping and protection of transactions with Blockchain.

In this chapter, we explore many of the domains, applications, architectures and related future technologies alongside the challenges that this field would face. While the main objective is to explore IoT's various facets, some of the most recent technological innovations, changing role of the Internet, and different IoT applications, considering Blockchain as a key enabler of security in future IoT, we have also explored the link between Blockchain and IoT.

We have divided the discussions in the chapter as follows: after the Introduction in Sect. 1, Sect. 2 talks about the application areas of IoT, Sect. 3 presents the IoT architecture. The IoT applications have been further explored in Sect. 4. Then, in Sect. 5, we explore Blockchain technology for IoT. Section 6 talks about IoT for Bangladesh, considering it as a representative developing country. This section is followed by Sect. 7 that talks about the real challenges for its implementation and growth in Bangladesh or similar setting. Also, the issue of Blockchain technology associated with this is discussed given this environment; Finally, Sect. 8 concludes the chapter.

2 Application Areas of Internet of Things

In this section, we explore the application areas of IoT. For easy navigation and reading, we have divided the subsections with the following headings: Internet of Smart Cities (IoSC), Internet of Smart Living (IoSL), Internet of Smart Health (IoSH), Internet of Smart Industry (IoSI), Internet of Smart Agriculture (IoSA), and The Network of Smart Things. While some of the applications and aspects may have commonalities among themselves or could be considered under more than one heading, this way of organizing the issues would be easier for the readers to grasp the concepts.

2.1 Internet of Smart Cities (IoSC)

2.1.1 Transportation Tracker

Transportation tracker [8] provides asset tracking, Geofencing (which is basically the use of GPS or RFID (Radio Frequency IDentification) technology to create a virtual geographic boundary, enabling software to trigger a response when a mobile device enters or leaves a particular area), Fleet management, vehicle's health maintenance, and so on. To illustrate the benefits of such application, let us take the case of traffic jam which no one basically likes. It wastes not only huge productive work hours but also consumes considerable amount of fuel or gas. For this purpose, traffic jam assistance technologies with transportation tracker could inform the driver about the fastest possible route to the destination and how much time it would be needed to go there. Also, such system could control or maintain the speed of a car based on the speed of the car(s) ahead (in the same lane) when it is a *slow moving* traffic congestion situation.

2.1.2 Traffic Management System with Green Environment

Traffic management system could be made more sophisticated with IoT technologies. For instance, sensors can be attached to the vehicle traffic signal lights or other road lights along the streets that would screen the air quality. These sensors could measure the level of CO, CO_2, NO, and NO_2 (or, gases that pollute the air) for a road and surrounding area and pass the readings to a local air quality administration center. Based on the reading, the administration center could combine other data about traffic conditions of various roads and then, all these data could be used for changing of signals and diverting some vehicles towards other alternative roads and bypasses. This could ensure a greener environment with better air quality for the area through which huge number of vehicles are passing. With today's computing capabilities and connected set of devices and mobility capability, such kind of reconciliation of the activity administration is possible. Figure 1 shows the concept.

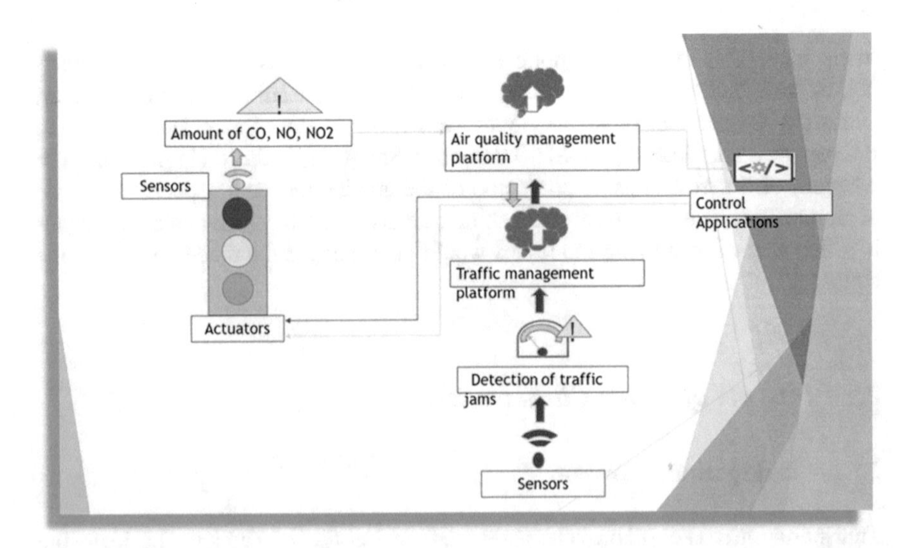

Fig. 1 Traffic management system

2.1.3 Smart Waste Management

Waste management is a big challenge in urban cities like Dhaka. Smart waste management is a framework that provides earlier information on the status of the waste container to the municipality. Consequently, this application can detect dry and wet waste, connects garbage machines, repack waste for better compaction as well. For instance, IoT infrastructure could set the platform for smooth functioning of garbage/waste management using CrAN (Crowd Associated Network) as described in Azad et al. [5].

2.1.4 Smart Parking

With the assistance of GPS, information from drivers' cell phones (or street surface sensors inserted in the ground on parking spaces), smart stopping arrangements can decide if the parking spaces are available or accessible and make an ongoing stopping map. At the point when the nearest parking space turns out to be free, drivers get a notice and utilize the guide on their telephone to discover a parking space quicker and simpler rather than aimlessly driving around.

2.1.5 Public Safety

IoT could be used for ensuring public safety in city areas. Information collected from acoustic sensors and CCTV (Closed-Circuit TeleVision) cameras could be recorded and monitored to prevent unwanted and/or criminal activities. With today's technologies, it is possible to distinguish the sound of a gunshot from any other sound. For instance, Boomerang gunfire locator developed by DARPA and BBN Technologies [9]. It is also possible to make the live-stream video feeds available to the public so that they could add extra eyes on important landmarks and places in a city. This could enable the police to stop potential culprits or effectively track them.

2.1.6 Utilities

Various types of utilities could be benefited and served by Internet of Things (IoT). Some utility related applications are:

Smart meters and charging: With a system of smart meters, countries can provide better service to their citizens. Smart meters can send information using the IT (Information Technology) infrastructure and it is possible to bill and charge precisely for the use of water, electricity, and gas. The idea of Smart grid IoT is also discussed in the recent days [10].

Assessing user behavior: A system of smart meters can empower organizations to accurately assess how their clients expend electricity and water. It can help make proper design for the supply grid. IoT again comes as a key aiding technology for this recent days [10, 11].

Remote monitoring and control: In connection with the previous two points, IoT based smart metering system can ensure better administration and distribution of utilities and greater remote monitoring. For instance, if it is possible to detect water leakage or electricity short circuit, the IoT based smart system can remotely turn off the supply to prevent mishap or at least in case of irregular usage pattern, warn the householder to check into the issue whether something has gone wrong.

2.1.7 Green and Clean Environment

IoT based systems can indeed help achieve green and clean environment. Various types of sensors can check water quality, the noise level in a city, pH level of soil, air quality, and so on. A warning could be issued if the continuous or periodic monitoring finds an irregular or toxic level of any of these and then, appropriate measures could be taken by the city corporation. This could make the concept of smart city a reality. Already, we have seen various works in this direction as reported in Al-Turjman et al. [12].

2.2 *Internet of Smart Living (IoSL)*

2.2.1 Smart Home Appliances

Home automation [13] with active assistance via the means of various electronic devices is already a known concept. Home automation system will monitor home appliances, entertainment system, lighting, climate, and so on. Smart home appliances are things that make one's life easier. People can give instructions by saying to a machine what to do and it obeys all of the instructions accordingly. Some real smart products are already available in the market. Let us talk about some of them briefly here:

Samsung smart refrigerator: Companies like Samsung represents Smart refrigerators with LCD (Liquid Crystal Display) screens that notifies about what is inside the fridge, ingredients that have to be purchased within a short time, food that is about to expire, planning and creating shopping lists, checking new recipes, and so on. The user can talk to the refrigerator's screen when hands are full, get entertainment with movie, music, television, and radio, etc.

Amazon echo: Amazon echo [14] is a famous smart speaker which is developed by Amazon. It is connected to a voice-controlled intelligent personal assistant named *Alexa*. It has several names "Echo", "Computer", "Alexa". From this device, users can set alarms, music, and stream podcasts, make a *to-do* list as well as provide information on road traffic and weather.

Ecobee: A Canadian home automation company launched Ecobee that is a thermostat used for residential and commercial area. Ecobee is controlled by using a friendly application available for Android, Apple Watch as well as iOS.
Smart home smart vent: Smart Vent by the Smart Zoning System can intelligently set room temperature using the smart home app. Often, some embedded programming is also used for devices like air coolers or air conditioners for the same.

2.2.2 Smart Weather Tracker Application and Device

Weather trackers are often installed on various handheld devices which could provide reading about outside climate conditions or forecast the weather. There are also weather tracking devices like indoor thermometer, wind speed gauge, outdoor thermometer, etc. These applications and weather tracking devices could inform the user about temperature, pressure, humidity, wind speed and rain levels with the ability to transmit data over long distances.

2.2.3 Smart Shopping

Nowadays, shopping has become smarter. Shopping malls like "Amazon Go" use "walk out technology" that can automatically detect products with prices using computer vision, sensor fusion, and deep learning algorithms. The stores are partially automated, with customers able to purchase products without being checked out by a cashier (i.e., a human operator) or using a self-checkout station. This would reduce the shipping time as well as billing time. Moreover, customers can pay through smartphones or by other electronic means.

2.3 Internet of Smart Health (IoSH)

2.3.1 Dental Applications

Today's Bluetooth associated toothbrush with smartphone application investigates the brushing uses. It also provides the conditions about tooth to the dentist.

2.3.2 Physical Activity Tracker

Remote sensors placed over the bedding detecting small movements, such as breathing and pulse and extensive movements caused by tossing and during sleep, can provide accessible information through an application on the smartphone. MSK and Medidata Cloud [15] developed an activity tracker that stores everyday information of a cancer patient. Moreover, it is necessary to wear the tracker before starting the treatment.

2.3.3 Smart Contact Lens

Smart contact lens technology is still under developed. It can detect various types of eye disorders like presbyopia as well as can measure diabetes of a patient through his/her tears and could save data on the smartphone.

2.4 Internet of Smart Industry (IoSI)

2.4.1 Dangerous and Hazardous Gases

Gas leakage detector can identify leakages and gas levels in the industrial environment; monitor the level of toxic gas as well as oxygen in chemical plants in order to provide safety to laborers and goods. In addition, it also monitors the oil, gas and water levels in cisterns and storage tanks.

2.4.2 Repair and Maintenance

By installing sensors in the equipment, equipment malfunction can be detected early and service maintenance can be scheduled automatically before the actual failure occurs.

2.5 Internet of Smart Agriculture (IoSA)

2.5.1 Animal Farming

Animal farming applications are used for area identification of animal grazing in open fields or to find out the location of the stables, and so on. Also, there are mechanisms to study air quality and ventilation in firms. Moreover, it is possible to detect the emission of harmful gases from the sediments or animal feces.

2.5.2 Monitoring Offspring

Sensors and other technologies nowadays could be used to monitor the health of the animals and the factors around it to guarantee a productive and healthy domesticated animal. Use of such technology could minimize the death of newborns of animals and domestic pets.

2.6 The Network of Smart Things

A smart city—as any Internet of Things (IoT) setting, utilizes advanced and sophisticated things/devices attached to sensors and actuators. The sensors could gather data and send those via some cloud or network to some administration and decision making center. The actuators could enable gadgets to act—adjust the lights, limit the stream of water to the pipe with spillage, and so on. Here, let us discuss a few interesting issues and applications in a smart city built with a network of smart things.

2.6.1 Gateway

Any IoT framework involves two sections—a significant portion that is consisted of gadgets and system hubs, and a cloud part. The information cannot simply be transmitted from one section of the setting to another section automatically. There must be some entry ways—field passages. Field doors would encourage information transmission when an event occurs, then some preprocessing may be done and the transmission would happen via some cloud but carried through some gateway (to go out of the source part of the network).

2.6.2 Big Data Warehouse

A big data warehouse is like an information vault. In contrast to information leaks (where information is kept in an unorganized manner), it contains organized and categorized information. Each piece of information is properly evaluated, categorized and kept in the databases. In addition, a data warehouse could also keep records of the logical data about the associated things, e.g., when sensors were introduced, or when the directions were sent to gadgets' actuators by control applications.

2.6.3 Smart Parking by Monitoring and Basic Analytics

One interesting smart city application could be a smart car parking facility on some ground or open field. There are some sensors that can read the dampness of the soil and then assist in parking on the relatively better spot in rainy season. The information provided by the sensors can be displayed on an electronic board when a vehicle enters the parking field and based on the suitability of the parking spot (where there is a plain grass field or soil), the driver could make a decision. This could be a smart parking zone with the assistance of sensing technologies. This is different than the parking technology with GPS that we have mentioned before.

2.6.4 Deep Analytics

Any kind of real implementation of IoT would generate a huge amount of information within short span of time. A smart city that uses the sophisticated technologies should be able to verify and make sense of the data that are collected. Indeed, often the sensor and other devices' input data are very complex and interconnected. Hence, machine learning (ML) and other types of factual investigations would be needed for processing such huge amount of data and information. Consequently, deep data analytics would be a critical operational issue for such smart city based on IoT. Then, that can help make better traffic light signaling decisions for instance or can be helpful for any other similar application in a smart city.

2.6.5 Smart Control

Control applications guarantee better computerization of smart city by sending directions to the actuators. Fundamentally, they "tell" the actuators what to do to comprehend a specific *undertaking*. There are rule-based and ML-based control applications. Principles for guideline based control applications are characterized physically, while ML-based control applications use models made by ML calculations. These models are distinguished depending on information examination; they are tried, affirmed and frequently refreshed.

After this discussion, let us talk about the architecture of IoT in the next section.

3 Internet of Things Architecture

The IoT can be considered as the third wave of the World Wide Web (WWW) after we have seen static site web pages and person to person communication based web. IoT could provide excellent connection to any kind of electronic device so that anytime, anywhere communication and data exchange is possible using the Internet Protocol (IP) or whatever may be employed in the coming future as the commonly used protocol. Though there are different viewpoints of the architecture of IoT, generally, we can consider three layers according to the work presented in Abdmeziem et al. [16]:

- Perception Layer
- Network Layer
- Application Layer.

From a different point of view of layering, an Assisting or Help Layer may be considered that would work in between Application Layer and System Layer of IoT. The assisting layer would help register devices and the process of distributed computing involved in the setting.

Perception layer helps perceive physical properties of things in an IoT setting. It could include several kinds of sensing technologies (RFID, NFC (Near-Field Communication), etc.) for the task of perception. While some objects are easily perceived or identified, some may be difficult to perceive. Hence, there could be additional microchips and identifying materials alongside using embedded intelligence for the devices which would work at this layer.

The Network layer is basically for the networking task and it is for transmitting data to the application layer for further processing. Also, communication between devices or host-to-host could be done at this layer. Network layer could use the regular wireless or wired networks or Local Area Networks (LANs) for the data transmission.

Application layer is the actual front-end of the IoT setting that would allow the users to interact with the devices or IoT objects or things. A particular IoT setting could be configured or used in certain ways through this layer. There could be a wide variety of applications like logistics management, location identification, safety management, intelligent transportation, logistics management, etc.

While the work in Abdmeziem et al. [16] presents three layers for the IoT operational architecture, the work in IoTSense [17] presents the following seven layers:

3.1 Layer 1: The Things Layer

This layer involves gadgets, sensors, and controllers. Associated gadgets are what empower the IoT setting. These gadgets incorporate cell phones; for example, advanced cells or tablets, miniaturized scale controller units and single-board PCs (Personal Computers). The associated gadgets are the genuine endpoint for the Internet of Things. There are sensors, other equipment; for example, implanted frameworks, RFID labels per user and others. The sensors empower the interconnection of the physical and virtual spaces enabling continuous data to be gathered and handled. The scaling down of equipment has enabled the sensors to be delivered in a lot smaller structures which are incorporated into various devices in the physical world. There are different kinds of sensors for various purposes. The sensors have the ability to take estimations and readings; for instance, temperature, pressure, air quality, magnetism, etc. At times, they may likewise have a level of memory, enabling them to record a specific number of readings.

3.2 Layer 2: Connectivity/Edge Computing Layer

This layer characterizes the different correspondence conventions and systems utilized for network and edge outlining. The IoT information is handled on the edge of the network. This layer deals with flow control, reliability issues, Quality of Service (QoS), and energy optimization. Also, it is responsible for cross-layer correspondence, whenever required. The IoT web-based interface is placed in this layer.

3.3 Layer 3: Global Infrastructure Layer

Layer 3 is the worldwide foundation layer, which is commonly actualized in cloud framework. The majority of the IoT arrangements are incorporated with cloud administrations. An extensive arrangement of coordinated administrations plays out the accompanying capacities; Gateway, Routing and Addressing, Network Capabilities, Transport Capabilities, Error location, and Correction. Additionally, it deals with message steering, distribution and buying. Numerous systems with different advancements and access conventions are expected to work with one another in a heterogeneous design. These systems can be of various types—private, open or half-breed models and they should work to help the correspondence necessities for transmission capacity and security.

3.4 Layer 4: Data Ingestion Layer

Layer 4 is the information ingestion layer, which incorporates Big data, purging, spilling and capacity of information. It deals with QoS Manager, Device Manager, Business Process Modeling, Business Process Execution, Authorization, Key Exchange and Management, Trust and Reputation, and Identity Management. In this layer, all activities require control and security for various applications.

3.5 Layer 5: Data Analysis Layer

In this layer, data are analyzed and mined. Various machine learning approaches may be used for that.

3.6 Layer 6: The Application Layer

Layer 6 is the application layer, which includes the custom applications that make use of the things' information. This layer is at the highest point of the design and is in charge of conveyance of different applications for various clients in the IoT. The applications can be from various industries, for example, industries related to production, co-ordination, retail, conditioning, human services, sustenance and medication, and so on. With the expanding development of RFID innovation and NFC technologies, various applications that are advancing will also be under the umbrella of IoT.

3.7 Layer 7: People and Process Layer

Layer 7 is the general population and process layer. It involves individuals, organizations, joint effort and basic leadership which depends on the data obtained from IoT environment. The items to be identified differ from physical moving articles; for example, people, vehicles, to ecological factors—for instance, temperature, or dampness. The spots to watch are structures, universities, streets, and so forth.

4 Internet of Things Applications—Survey and Opportunities

According to McKinsey—an American worldwide management consulting firm, the economic value generated by IoT could reach US$11.1 trillion a year by 2025 [17]. While predictions can be accurate or not, the potential usage fields are huge and we are already witnessing the penetration of electronic gadgets in our daily life. We felt that there is a need to do some survey to explore the opportunities and current trend of this field in different parts of the globe so that these data could provide important direction for broadening the capability of IoT and associated technologies and concepts. Hence, to find out IoT application and domains in a better way, we performed a survey through which we wanted to know how people think about different IoT applications. We also collected their opinions about those applications.

We conducted a survey using Google in which a total of 200 people participated. We took the responses from the first 200 participants. The responders represented a wide range of ages, genders, and geographical locations. The results show that among the *technology-aware* people, around 74% are using some type of IoT application in their everyday life. Among them, 71% of were Males and 29% were Females. 81% of them were Students and 19% were Job Holders. Among these IoT application users, 75.9% were in age range 20–30, 20.9% were in the age range 30–40, and 3.2% were in the age range 40–50. As for the geographical diversity, 55% people were from

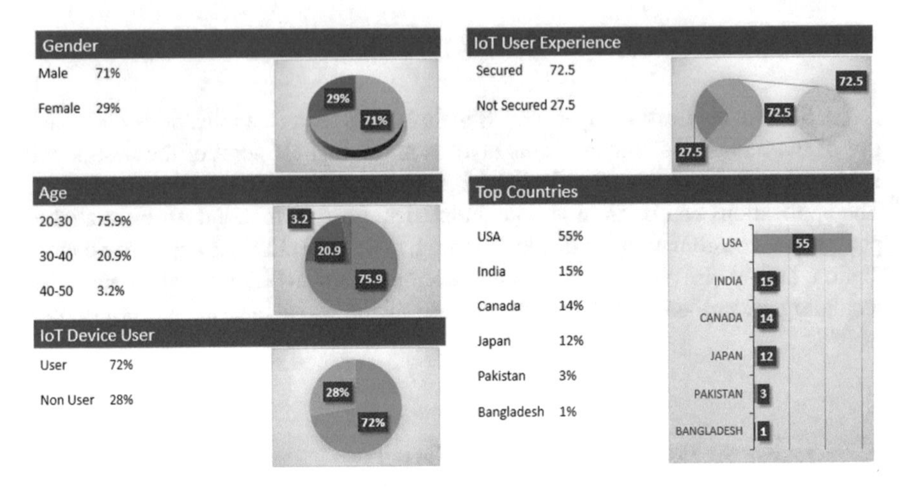

Fig. 2 Survey results regarding IoT applications

USA, 15% people were from India, 14% people were from Canada, 12% people were from Japan, 3% people were from Pakistan, 1% people were from Bangladesh. As it could be understood, the actual number of participants from Bangladesh for this online survey is too low; however, talking to people; we have understood their interest in IoT and its applications. It is difficult to motivate them to participate in the survey online as they may find some other tasks about daily life to be given more priority given the lifestyle or kind of hardship in this developing country's setting. However, general interest has been recorded. Figure 2 shows the results.

The survey shows that the people from a developed country like the USA are using IoT applications more than those in other developing countries like Bangladesh or Pakistan (as would be expected). Especially, in Bangladesh like setting, smart watch, smart TV, smart refrigerator or similar items and applications are being used. The initial introduction to IoT based system was in this country through services like vehicle tracking and smart electricity metering that started around a decade ago from the time of writing this chapter. Since then, there has been steady and somewhat silent growth in this arena, while in the recent years, it is receiving relatively higher focus considering the extensive range of potentials in this area.

As we have observed, the people of Bangladesh are highly interested in IoT applications in different sectors. This is understood from the people who we have talked to even though many of them did not actually participate in the online survey spending some time. Smart Alarm System, Smart Home Appliance, Smart Vehicle System, Smart Home, and Smart Mobile Tracker and similar applications are now in high demand in this country. In fact, from the business perspective, people's participation is a key factor for any technological field. Given the current state of the affairs, IoT has a promising future with various smart and essential applications that can help the daily life of the people in Bangladesh. In the past, we saw how fast mobile phone technology grew among these people and quickly, the operators got millions of subscribers. Any new technology related to IoT is similarly expected to hit the market with a very smooth pace and growth.

4.1 IoT Solutions

IoT will open new opportunities for administration and business to help the corresponding requests of possibly many billions of gadgets in the future after the real patterns that are being seen today with the use of RFID tags or so. We have to keep an eye on the use of such ultra-minimal effort labels while the data are unified on information servers overseen by different types of managerial and administrative units. Such use of easy-to-use electronic labels could expand in the coming days but the growth would need extra memory in the devices and detection capabilities. There may be too many connected devices and equipment around us participating in such auto-detection, data processing, and the decision-making process.

One of the most prominent advancements of RFID technology is an HF RFID (High-Frequency RFID) or NFC (Near-Field Communication) [18]. NFC empowers online stores to give clients an increasingly helpful online installment choice with an included security layer. Using check cards for NFC chips, clients can pay online with a solitary tap against an NFC installment station. This makes the process faster and easier to handle a huge number of people in the payment counters. One advantage of this technology is that there is no possible way for a rogue entity to extract information through the assistance of an attractive strip. This is because the NFC-driven installment framework does not include the card swiping process (which is a problem for swiping-based technologies). This online installment framework has been embraced by Google, a tech giant which prompted the ascent of Google Wallet. Though the method is attractive, a negative point for this is that it is confined by the number of retail locations that have NFC stations and telephones that should help in the process.

In fact, RFID technology overall has the potential to redefine almost every industrial sector as it helps customers have sufficient access to the relevant information about every online retailer's business offerings. RFID devices also keep the works well with constant customer engagement as it notifies them about every single update on a regular basis. These RFID tags also allow customers to keep track of their online orders and help retailers give relevant updates regarding their orders. RFID Tracking Mobile App empowers the users for asset tracking to perform tasks; for example, get, move, and execute physical inventories immediately with the assistance of an RFID-handheld per user. Bluetooth technology can be used in the process. RFID technology may also replace barcode scanning technology in the near future. Again, the HF RFID technology can be used in libraries to enable NFC communications. There are in fact, too many potential applications and all these would contribute to the IoT's development.

In such a setting, all these technologies would generate a huge amount of data which need to be conveyed and brought together to information servers or data centers alongside the electronic labels. In the coming days, we may see more functionalities for such electronic labels. In fact, data could be made effective yet lightweight for faster processing. What needs to be ensured is the proper level of security and effective synchronization of all the related and sequential data. Indeed, sophisticated gadgets with brilliant frameworks and fast data generation and processing capability would offer a high degree of knowledge about the environment where the gadgets would run and that would make sure self-ruling capability is in an IoT application enabling a new type of system administration.

4.2 Intelligent IoT Services and Platforms

Google Cloud IoT (GCIoT) is a complete set of tools to connect, process, store, and analyze data both at the edge and in the cloud [19, 20]. Using this platform, it is possible to run IoT solutions with machine learning capabilities, both locally and in the cloud. This is one of the notable advancements and platforms of what the future IoT related cloud service could look like. Figure 3 shows the Google Cloud IoT core architecture (adopted and modified from Sumit [19]).

Utilizing such a platform, it is possible to do various things like:

Accelerating business agility and decision-making with IoT data: It is possible to increase continuous business experience from universally scattered gadgets, at the edge or in the cloud, with far-reaching administration. So, essentially, real-time business insights can be gained from globally dispersed devices. Device data captured by Cloud IoT Core would get published to Cloud Pub/Sub for downstream analytics. Necessary analysis could be made using Google BigQuery or machine learning can be applied to the Cloud Machine Learning Engine. In Google Data Studio, the data could also be visualized.

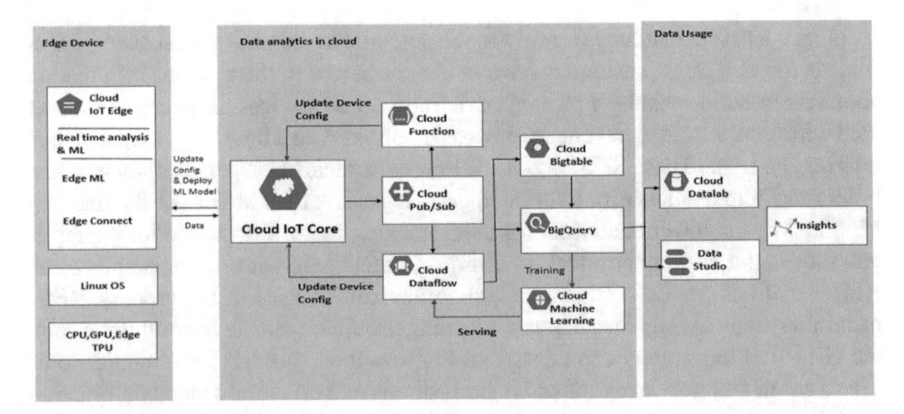

Fig. 3 Google Cloud Internet of Things architecture (adopted and modified from Sumit [19])

Conveying Artificial Intelligence (AI) to the edge: AI related techniques and capabilities could be implemented into the edge devices.
Enhancing Operational Effectiveness: For enhancing operational effectiveness, GCIoT could really help. It could allow the user to effectively monitor how the gadgets work, oversee worldwide resources, and complete firmware reports on GCIoT.
Enhancing the IoT solution with location intelligence: This could be an excellent way to make one's business *location-aware* with the power of Google Maps. The business owner can visualize where the assets are located in real-time, where they have traveled, and how often they have moved. Whether IoT assets are indoors, in remote areas, or distributed across hundreds of cities, the owner/user can track them with precision.

The impact of IoT is such today that there is not a single industry that has not faced some type of IoT application or potential IoT related issue at some point. Initially, some companies could not realize the actual impact of IoT and what would be in the coming days but it is good to see that many have already started understanding the significance of it and are taking appropriate step(s) to adopt IoT tools and applications to stay relevant to the futuristic heavily inter-connected market and global trend. In practice, nowadays we see that the retail organizations are exhorting intense efforts in IoT as they comprehend the significance of information-driven environment coming ahead and also to enhance the client experience and client encounter process. On the other hand, clients appreciate the new systems and methods that are set up with the aid of IoT technologies. In the recent years, billions of sensors and sensing devices have been used in many sectors and information collected by all these devices would be of huge size.

IoT is reshaping healthcare sector as well. Wearable technology to monitor one's physical condition at anytime and anywhere is very common now. Sensors can collect data and at the same time, the data can be visible to doctors who could take appropriate measures based on the readings. In this way, doctors can closely monitor crucial patients even from a distance [21, 22].

The manufacturing industry is also making use of smart machines to improve the overall manufacturing process and to produce better goods. In fact, we see that IoT has something to offer to everyone using technology and the Internet or such cyberspace—it has completely changed the way that the people used to do business, socialize, and have fun. IoT is indeed a game-changer and there is enough potential for professionals to innovate new applications and ways of operations.

5 Blockchain Technology for IoT

It is understood that IoT setting would produce tremendous volume of data—transactions and information exchange between participating nodes/entities would be of very high number. Secure and reliable operations for all these would be a great challenge but Blockchain seems to offer some solution for tracking and keeping record of such transactions and data transfers.

Blockchain is basically the accumulation of a series of '*blocks*' where each block stores data involved in a transaction When the transaction occurs, it is stored in a block and added to the '*chain*'. These blocks form a distributed database to store valuable information and thus, form a Blockchain. Each of the participants in the Blockchain has access to the same database because it contains shared databases, which are cryptographically secure to protect the data. For adding any new block in the chain, each of the participants needs to approve and therefore, the data in Blockchain would be technically *immutable*. Imagine for approval of someone's job, each member of a recruitment committee must vote positive. Otherwise, confirmation would not be done. Hence, in this technology, any intruder will not be able to inject into or delete any data from the Blockchain. In terms of data security, the common cyber attacks like Denial-of-Service (DoS) attack and many other types do not have any significant effect if the data are stored using the Blockchain. Therefore, this technology is being embraced at a rapid pace across a number of application domains [4] including IoT. The problem however remains with the fact that both Blockchain and IoT are going through lots of experimentations, conceptualizations, and imaginations. Also, the notion of traditional centralized system would be largely diminished when IoT comes into action in real-life activities. This may make the case even more unpredictable. As in the security community, we say that there is no hundred percent guarantee of security for any system and hence, whenever the *online* or *cyber* activities are involved, there would also be issues that may be negative or risky for IoT even if Blockchain is used for confidential operations. Another issue is that software vulnerabilities may allow ways for the attackers to get into the system and manipulate it.

In a recent work, Ahmed and Pathan [4] show that Blockchain cannot be fully trusted because, there are multiple recent incidents of security breaches in this technology or the technologies that it relies upon. Especially, in the context of smart contracts, web application or software vulnerabilities are of prime concern when Blockchain technology is considered. Smart contracts are basically the digital version (computer code) of traditional economic contracts among different parties involved in any trade or business. Unlike the regular contracts, the Blockchain-enabled smart contracts do not require any intermediaries to ensure the conditions to be fulfilled. Vulnerable smart contracts can simply spoil the environment of free data exchange in a somewhat friendly setting of IoT. Unless extensive research is done, some more time is given, and Blockchain is proven to really add benefit to the IoT environment, it would be wise to use other known or established security solutions for the protection of transactions and data exchanges in IoT.

6 IoT for Bangladesh as a Model Developing Country

IoT is mainly built upon three major components. To comprehend the status of IoT in a country like Bangladesh as a sample developing country, it is essential to analyze the conditions of these three fundamental building blocks. Here, we discuss each of them.

Equipment: More explicitly, these are sensors, which measure and gather related data from the source devices. In Bangladeshi market, most of these devices are sold for commercial purposes and privately owned gadgets are relatively less in amount. Sensors are often used for buildings, roadsides, door opening, and closing, etc. but personal information carrying and data-providing sensors would still be small in quantity till the time of writing this chapter. However, there is an increasing trend of getting used to with sensor-based electronic gadgets, especially among the youth.
Availability: Availability of various communications technologies is a concern. Given the current status, versatile wireless systems like 2G, 3G, 4G are supported as of today with the plan of inclusion of future technologies for mobile communications. Also, in many places WiFi and other wireless connection supports are available. Internet connectivity throughout the country is well established and is supported by many ISPs (Internet Service Providers) and mobile operators.
Software: Software involves the applications for stockpiling, preparing, examining client interfacing for significant information harnessing. With exceedingly able programmers' community and IT (Information Technology) professionals in Bangladesh, most of the software are developed locally. It is a promising sector of Bangladesh as these people could well-develop the required application software for IoT's support.

6.1 IoT Development Areas in Bangladesh

While IoT-based applications in the country have so far remained as mainly thoughts and visions around different areas of administrations, the potential development areas are indeed many, including horticulture, security and reconnaissance, smart home, smart city, automation, medical services and healthcare, and so on. In the recent days, there is also a huge interest shown both by individuals and various companies. For any real development of IoT in this status, many of the issues would come into play. Here, we discuss some of the noteworthy issues.

Widespread industry: Developing local industries for self-supportability is very important for IoT's administration and associated operations. There are three issues that an IoT industry needs to handle: equipment (neighborhood producing and collecting), network (inclusion of the known and most recent communications protocols and techniques like Narrowband IoT, LoRA—long-range wireless communication), and programming (information facilitating and tweaked apps).

Acquaintance: Creating awareness and acquaintance among clients is very crucial to expanding the notion of IoT and the reception of its benefits.
Building skill: Required mastery (specialized, showcasing, regulatory) in development among people working in this industry is critical. Hence, building skill is a must.
Strategies: Formulating applicable government arrangements and administrative rules to nurture IoT is crucial. The simple first directive with respect to IoT (draft policy) was distributed by Bangladesh Telecommunication Regulatory Commission (BTRC) in April 2018 [23], which generally covers the laws and rules on IoT gadget import. There are a few different viewpoints which require far-reaching directives; for example, security and protection in IoT-based administration, upgrading of SIM (Subscriber Identity Module) based IoT administration, compatibility of various types of telecom administrators, duties, and obligations with respect to the gadget and information charges, and so on.
Dispersion channel: Developing nationwide dissemination and *after-deal* policies could bolster the dispersion channel for IoT devices and thus help the IoT administration. Though at present, various sectors are contemplating the use of IoT for their own purposes (main usages as mentioned in the BTRC directive are; Smart Building, Smart Grids, Telecare, Industry Automation, Water Management, Smart Agriculture, Intelligent Transport System, Waste Management, Smart Parking, Environment Management, and Smart Urban Lighting), to make use of the market in the best manner, legitimate deals and administration channels should be opened and regulated.
Industry-Academic cooperation and research: Academic help for creative brilliance, i.e., the industry-and-academic joint effort is needed. An extensive number of academic works are now focusing on IoT related research issues, which is a good sign. Alongside the researchers and academics in the developed countries, in Bangladesh, various universities are also encouraging similar research works. Once some well thought-through solutions and ideas are available, those could be publicized among the general mass.

One of the positive points for the development and adoption of IoT at a fast speed in Bangladesh is its huge population. Given the history of quick spreading of cellular networks and any latest technology and its acceptability among the common populace, it is possible to be hopeful about IoT's widespread use in this country. Also, a large force of programmers, IT professionals could easily contribute to the growth of IoT.

6.2 *Major IoT Opportunities in Future Bangladesh*

As noted before, there is at least some directive today from the government of Bangladesh about IoT and related applications. Besides those mentioned in the

BTRC document, here we discuss a few other application areas that can grow fast in Bangladesh given today's technology trend.

Smart Alarm System: Smart alarm system is a special type of security alarm system that allows the user to monitor home as well as office security. A user can check who is outside of the door, whether someone is knocking or not and so on. In addition, a user can lock and open the door by using mobile phone.
Mobile Phone Tracking: Mobile phone tracking is another application of the Internet of Things. A user can know the location of the mobile phone by this application. It will help a lot if someone's mobile phone is lost.
IoT Machine Learning: Machine learning could be really helpful with IoT for solving various types of system related problems like traffic management system and population or crowd management, queue management while selling bus or train tickets, and so on. In Bangladesh, this area has huge potential.
Big Data Analytics: With the huge population in Bangladesh, the amount of data generated is also huge (for almost anything). For any system, it would be really difficult to handle such a large volume of data. IoT, if really comes into force in daily life affairs in Bangladesh, big data analytics would be also very much relevant to the data processing tasks when so many devices would be connected.

7 Challenges for IoT in Bangladesh

The term, 'IoT' is still mostly meaningful for the developed countries. In this type of developing country, the challenges for the real-life implementation of IoT would be great in number or at least of great negative strength, even if few.

In the year 1989, innovation showed a remarkable leap towards advanced life and communication started among human beings at a distance with the then cutting-edge technology developed by scientist Tim Berners-Lee which was termed World Wide Web (WWW) and it quickly spread around the globe afterwards. Nowadays, the Internet is no separate issue from our daily life when we use the least level of technology or distant communication and information sharing. It should be noted that Internet is basically a globe-covering network of networks while the Web, or as termed formally, World Wide Web (WWW) is a collection of information which is accessed via the Internet. Since that time of early development of all these, till today Bangladesh has welcomed (or, welcomes) the web related innovations and like other developing countries, established own administrative measures. Here, however, come the issues and challenges that are unique given the demography, politics, and other rules and laws established in Bangladesh.

IoT has legitimately entered Bangladesh with a directive issued by Bangladesh Telecommunication Regulatory Commission (BTRC). However, in the past, there were people in the decision-making bodies who disliked the idea of information sharing or considered the expansion of even the Internet or Internet-like infrastructure some kind of security threat for the country. This is because the political arena of this country was very volatile from time to time and there were several irregularities in elections or grabbing power or not following all internationally accepted norms of holding ruling positions. Political clashes and disputes disrupted the growth of the Internet and later technologies for quite long number of years. With the vision of Digital Bangladesh [24], the issues have become a bit more congenial but there are often many restrictions in using these types of platforms. Often, BTRC could block the services like Facebook, Twitter or other social media to restrict the flow of information and there have been several recent cases. Also, some news channels may also be blocked or made inaccessible. Given this setting, for IoT, the real challenge lies not on the issue of acceptability among the common mass but mainly on the political setting of this country. After all, a stable country ruled by laws as are formed could enable smooth and proper growth of IoT and its associated activities.

Apart from the political challenges that may really affect any IoT decision even in the future, there is some type of obstacle in accessing the most advanced technologies in the areas just outside the major cities in this country. Electricity supply is another serious challenge as all these IoT devices must need power or electricity to run, even if they can be on batteries from time to time instead of constant electricity use from direct supply. Often, during summer days, when heat-wave starts, there are frequent power outages in various places in this country and the power demand is way higher than the current total generation and distribution capacity. Even the furthest prediction does not show a sign of complete relief out of this situation for the entire country but some parts may be well-supplied with constant electricity within few years from now. This puts a serious challenge in seeing IoT a reality outside the posh areas or even outside the major cities and places in Bangladesh.

Regarding the use of Blockchain for IoT and other domains, recent news that could be cited is that the Bangladesh government is planning to use money from its US $208 million IT project fund to send graduates for Blockchain training in Japan and India [25]. Till the time of writing this chapter, the government is yet to have a clear policy on Blockchain though learning the technology is encouraged from various quarters. It should be mentioned that Bitcoin technology for which the concept of Blockchain showed promise (mainly) to be used as a public ledger to keep track of all the transactions, is not welcome in Bangladesh. In fact, as per the report by Mahbub and Rahman [26], Bangladesh is one of only 6 countries in the world that are considered "*hostile*" to Bitcoin. Hence, any other domain except Bitcoin is fine for Blockchain use in Bangladesh at this stage.

Previously, an IDC (International Data Corporation) forecast showed worldwide spending on the IoT to reach $772 Billion in 2018 (which was more or less in reality). The spending would increase as the time passes by. The innovation indeed requires huge spending to produce real-working gadgets, let alone the theoretical expansion and research in regular academia and research labs. As we have investigated the area, even with BDT 500 billion of interests in IT area of Bangladesh, there is no explicit information to discover the total sum of the employed resources for IoT; however our an assumption could be made based on the currently accessible IoT items and administrative facilities and this could be under 5% of the BDT 500 billion sum. Whatever spending is done, ultimately, different public and private segments and the government should allow each other some space for the growth of own IoT related activities. Such collaboration and cooperation often remain a serious challenge in Bangladeshi setting.

8 Conclusions

Technology is necessary to achieve the ubiquitous network society that we expect for the future. With the tremendous advancements of electronics and IT, we have already entered some type of mature stage. Based on the current platform, the IoT is growing at a rapid pace. With the promise of easing daily life and governance of people and sharing information, it has attracted too many sectors and people in the globe. Yet, the harmful sides remain and without tackling the negative sides, the hype may not really deliver the benefits at the end. With the growing economy of Bangladesh, IoT is also a technology to get benefit from and our study showed that the people are generally ready for accepting it for their daily life. When the security and privacy issues would be better shaped and the government would be better convinced with the benefits manifested in other developed countries, perhaps, wide-scale implementation of IoT in Bangladesh would not take any long time. Already, all technologies that we see in other developed countries are more or less available in Bangladesh at least at a restricted scale. The willingness of the people to welcome new technologies and the vision of a digitally capable society are the positive points for successful IoT-based setting in Bangladesh (or, any other developing country in the similar status) in the coming days.

References

1. Pathan, A.-S.K., Saeed, R.A., Feki, M.A., Tran, N.H.: Guest editorial: Special issue on integration of IoT with future internet. J. Internet Technol. **15**(2), 145–147 (2014)
2. Pathan, A.-S.K., Fadlullah, Z.M., Choudhury, S., Guerroumi, M.: Internet of Things for smart living. Wireless Netw. (2019). https://doi.org/10.1007/s11276-019-01970-3
3. Guinard, D., Trifa, V., Pham, T., Liechti, O.: Towards physical mashups in the web of things. In: INSS'09 Proceedings of the 6th International Conference on Networked Sensing Systems, pp. 196–199, Pennsylvania, USA, 17–19 June 2009
4. Ahmed, M., Pathan, A.-S.K.: The Blockchain: can it be trusted? IEEE Computer, 2019 (to appear)
5. Azad, S., Rahman, A., Asyhari, A.T., Pathan, A.-S.K.: Crowd associated network: Exploiting over smart garbage management system. IEEE Commun. Mag. **55**(7), 186–192 (2017)
6. Jamoussi, B.: IoT prospects of worldwide development and current global circumstances. Slides available at https://www.itu.int/en/ITU-T/techwatch/Documents/1010-B_Jamoussi_IoT.pdf. Accessed 30 Apr 2019
7. Cisco Visual Networking Index: Forecast and Trends, 2017–2022, White Paper, Feb. 2019. Available at https://www.cisco.com/c/en/us/solutions/collateral/service-provider/visual-networking-index-vni/white-paper-c11-741490.pdf. Accessed 30 Apr 2019

8. Kalid, K.S., Rosli, N.: The design of a schoolchildren identification and transportation tracking system. In: 2017 International Conference on Research and Innovation in Information Systems (ICRIIS), 16–17 July 2017. https://doi.org/10.1109/icriis.2017.8002454
9. Boomerang shooter detection technology. Available at http://milcom-security.com/wp-content/uploads/BoomerangGeneral-102010-5.pdf. Accessed: 2 May 2019
10. Fadlullah, Z.M., Pathan, A.-S.K., Singh, K.: Smart grid Internet of Things. Guest editorial of the special issue of ACM/Springer Mob. Netw. Appl. **23**(4), 879–880 (2018). https://doi.org/10.1007/s11036-017-0954-2
11. Zeadally, S., Pathan, A.-S.K., Alcaraz, C., Badra, M.: Towards privacy protection in smart grid. Wireless Pers. Commun. **73**(1), 23–50 (Nov 2013) https://doi.org/10.1007/s11277-012-0939-1
12. Al-Turjman, F., Kama, A., Rehmani, M.H., Radwan, A., Pathan, A.-S.K.: The Green Internet of Things (G-IoT). Wireless Commun. Mob. Comput. **2019**, 6059343 (2019) https://doi.org/10.1155/2019/6059343
13. Hamdan, O., Shanableh, H., Zaki, I., Al-Ali, A.R., Shanableh, T:. IoT-based interactive dual mode smart home automation. In: 2019 IEEE International Conference on Consumer Electronics (ICCE), 11–13 Jan 2019. https://doi.org/10.1109/icce.2019.8661935
14. Ludlow, D.: Google Home vs Amazon Echo: Which is the best smart speaker?, 22 May 2019. Available at https://www.trustedreviews.com/news/google-home-vs-amazon-echo-2945424. Accessed 1 June 2019
15. AI and IoT in healthcare—current applications and possibilities. Available at https://emerj.com/ai-sector-overviews/ai-and-iot-in-healthcare-current-applications-and-possibilities/. Accessed 11 May 2019
16. Abdmeziem, M.R., Tandjaoui, D., Romdhani, I.: Architecting the Internet of Things: state of the art. In: Robots and Sensor Clouds, pp. 55–75. Springer, Cham (2015)
17. The layers of IoT—IoTSense. 2019. Available at http://www.iotsense.io/blog/the-layers-of-iot/?fbclid=IwAR3gxrGjfkGTH1u2KX7rWC0LAwJLD696qhd7m_4wWlE7q3IMPtUQhCz9Brg. Accessed 16 May 2019
18. Washiro, T. HF RFID transponder with capacitive coupling. In: 2017 IEEE International Conference on RFID Technology and Application (RFID-TA), 20–22 Sep 2017
19. Sumit. Overview of Google Cloud IoT core. Available at: https://www.systemadminworld.com/2018/11/overview-of-google-cloud-iot-core.html. Accessed 28 May 2019
20. Google Cloud IoT. Available at https://cloud.google.com/solutions/iot/. 1–23. Accessed 28 May 2019
21. Khan, R.A., Pathan, A.-S.K.: The state of the art wireless body area sensor networks—a survey. Int. J. Distrib. Sens. Netw. **14**(4) (2018)
22. Fadlullah, Z.M., Pathan, A.-S.K., Gacanin, H.: On delay-sensitive healthcare data analytics at the network edge based on deep learning. In: The 14th International Wireless Communications and Mobile Computing Conference (IWCMC 2018), pp. 388–393, Limassol, Cyprus, 25–29 June 2018
23. BTRC allows IoT solutions in Bangladesh. The Daily Star, 24 Apr 2018, Available at https://www.thedailystar.net/country/bangladesh-bd-telecommunication-regulatory-commission-btrc-allows-internet-of-things-iot-solutions-technology-1567198. Accessed 5 June 2019
24. IDC forecasts worldwide spending on the Internet of Things to reach $772 billion in 2018. Available at https://www.idc.com/getdoc.jsp?containerId=prUS43295217&pageType=PRINTFRIENDLY. Accessed 8 June 2019
25. Partz, H.: Bangladesh to use IT fund to bankroll Blockchain education for graduates. Available at https://cointelegraph.com/news/bangladesh-to-use-it-fund-to-bankroll-blockchain-education-for-graduates. Accessed 1 Sep 2019
26. Mahbub, S., Rahman, R.: Legality of bitcoin in Bangladesh, 10 July 2018. Available at https://www.thedailystar.net/law-our-rights/law-analysis/bitcoin-legality-in-bangladesh-bank-1602583. Accessed 1 Sep 2019

Correction to: Legal Ramifications of Blockchain Technology

Akinyemi Omololu Akinrotimi

Correction to:
Chapter "Legal Ramifications of Blockchain Technology" in: M. A. Khan et al. (eds.), *Decentralised Internet of Things*, Studies in Big Data 71, https://doi.org/10.1007/978-3-030-38677-1_10

In the original version of the book, the following belated correction has been incorporated: The order of the author's name has been changed from Akinrotimi Akinyemi Omololu to Akinyemi Omololu Akinrotimi in chapter "Legal Ramifications of Blockchain Technology". The chapter and book have been updated with the changes.

The updated version of this chapter can be found at
https://doi.org/10.1007/978-3-030-38677-1_10

M. A. Khan et al. (eds.), *Decentralised Internet of Things*, Studies in Big Data 71,
https://doi.org/10.1007/978-3-030-38677-1_12

Printed in the United States
By Bookmasters